The Entomology of Indigenous and Naturalized Systems in Agriculture

The Entomology of Indigenous and Naturalized Systems in Agriculture

Westview Studies in Insect Biology
Michael D. Breed, Series Editor

Fire Ants and Leaf-Cutting Ants: Biology and Management, edited by Clifford S. Lofgren and Robert K. Vander Meer

Management of Pests and Pesticides: Farmers' Perceptions and Practices, edited by Joyce Tait and Banpot Napompeth

Integrated Pest Management on Rangeland: A Shortgrass Prairie Perspective, edited by John L. Capinera

Interindividual Behavioral Variability in Social Insects, edited by Robert L. Jeanne

The Entomology of Indigenous and Naturalized Systems in Agriculture, edited by Marvin K. Harris and Charlie E. Rogers

The "African" Honey Bee, edited by David J.C. Fletcher and Michael D. Breed (forthcoming)

The Genetics of Social Evolution, edited by Robert E. Page and Michael D. Breed (forthcoming)

The Entomology of Indigenous and Naturalized Systems in Agriculture

EDITED BY

Marvin K. Harris
and Charlie E. Rogers

Routledge
Taylor & Francis Group

NEW YORK AND LONDON

First published in paperback 2024

First published 1988 by Westview Press, Inc.

Published 2019 by Routledge
605 Third Avenue, New York, NY 10158

and by Routledge
4 Park Square, Milton Park, Abingdon, Oxon OX14 4RN

Routledge is an imprint of the Taylor & Francis Group, an informa business

© 1988, 2019, 2024 Taylor & Francis

Library of Congress Cataloging in Publication Data
The Entomology of indigenous and naturalized systems in agriculture /
  edited by Marvin K. Harris and Charles E. Rogers.
      p. cm.—(Westview studies in insect biology)
    ISBN 0-8133-7621-1
    1. Insect pests. 2. Insect-plant relationships. 3. Agricultural
ecology. 4. Crops—Origin. 5. Cropping systems. I. Harris, M. K.
(Marvin K.) II. Rogers, C. E. (Charlie E.), 1938– . III. Series.
SB931.E57 1988
632'.7—dc19   88-10747
                                                                          CIP

Publisher's Note
The publisher has gone to great lengths to ensure the quality of this reprint but points out that some imperfections in the original copies may be apparent.

ISBN: 978-0-367-29179-2 (hbk)
ISBN: 978-0-367-30725-7 (pbk)
ISBN: 978-0-429-31054-6 (ebk)

DOI: 10.1201/9780429310546

Dedication

The Annual Robert H. Nelson Symposium on Crop Protection Entomology was established in 1985 by the Entomological Society of America (ESA) on behalf of Section F to recognize the past Executive Secretary (1955–1968) and President (1971) of ESA. This book derives from the first two Symposia, also co-sponsored by Section C, and is dedicated to Robert H. Nelson.

He was born in Nebraska in 1903, received the B.S. and M.S. from the University of Nebraska and pursued further studies at Iowa State College and The Ohio State University. Bob began his professional career in 1931 with the USDA and from 1946 to 1954 was responsible for the conduct, and ultimately the coordination, of the USDA's pesticide screening program for arthropods. He dedicated the remainder of his career to the development of ESA and characteristically his Presidential Address (Bull. ESA 18[1]:2–6) recognized the 18 ESA Presidents that preceded him. Sustainable progress is achieved by providing the foundation and mechanisms for ordered change and Robert H. Nelson tended those fires well in the service of his fellow entomologists. We recognize his worthy and essential work in this small way and wish him many more happy years.

Contents

Preface

Reports on basic entomological research themes have newly emphasized living systems of note to agriculture. Agricultural systems dominate our planet and the organisms associated with the agroecosystems provide unique opportunities to further our understanding of biological processes because of the wide range of environments, from the pristine to the intensively managed, that are naturally available for study. This variation is a blessing because it provides an easily accessible basis for comparisons, and a curse because the hand of man can create conditions that are so removed from their natural environmental settings that useful parallels cannot be easily investigated.

The purpose of this book is to highlight some agriculturally important plants and their associated arthropod complexes with a biological, as well as an agricultural, perspective. The authors were charged with relating the latter to the former as much as possible by emphasizing how the wild plant interacted with arthropods prior to, as well as after, plant domestication and how this knowledge could be used to further progress in solving problems facing biology and agriculture.

The vignettes that follow serve as a beginning for narrowing the wide chasm currently dividing these perspectives, given agricultural scientists inspired to investigate biological processes and biological scientists inspired to investigate agricultural systems. The intellectual and material well-being of mankind will be influenced by how well we meet this challenge.

Contributors

John N. All, Professor of Entomology, University of Georgia, Athens, Georgia 30602.

Ann Amis, Research Entomologist, ARS/USDA Southeastern Fruit and Tree Research Laboratory, P. O. Box 87, Byron, Georgia 31008.

James D. Dutcher, Associate Professor of Entomology, Coastal Plain Experiment Station, Tifton, Georgia 31793.

Stan Finch, Professor, Entomology Section, National Vegetable Research Station, Wellesbourne, Warwicks, England CV 35 9EF.

Frank E. Gilstrap, Professor of Entomology, Texas A&M University, College Station, Texas 77843.

Marvin K. Harris, Professor of Entomology, Texas A&M University, College Station, Texas 77843.

E. A. Heinrichs, Head, Department of Entomology, Louisiana State University, Baton Rouge, Louisiana 70803.

Dan L. Horton, Extension Entomologist, University of Georgia, Athens, Georgia 30602.

James E. Litsinger, Entomologist, International Rice Research Institute, P. O. Box 933, Manila, Phillippines.

Michael E. Loevinsohn, Program Officer, International Research and Development Centre, P. O. Box 8500, Ottawa, Canada, K1 G3H9.

Kathryn C. McGiffen, Illinois Natural History Survey, Natural Resources Bldg., 607 E. Peabody Drive, Champaign, Illinois 61820.

Jerry A. Payne, Research Entomologist, ARS/USDA Southeastern Fruit and Tree Nut Research Laboratory, P. O. Box 87, Byron, Georgia 31008.

Charlie E. Rogers, Laboratory Director, Insect Biology and Population Management Research Laboratory, ARS/USDA, Tifton, Georgia 31793.

Carl H. Shanks, Jr., Superintendent and Entomologist, Southwestern Washington Research Unit, Washington State University, 1919 N. E. 78th St., Vancouver, Washington, 98665.
Thomas M. Sjulin, Small Fruit Breeder, Western Washington Research and Extension Center, Puyallup, Wash. 98371
George L. Teetes, Professor of Entomology, Texas A&M University, College Station, Texas 77943.
Michael J. Way, Professor of Applied Zoology, Imperial College, Silwood Park, Ascot, Berks. SL5 7PY, United Kingdom.

1

Entomology of Indigenous *Helianthus* Species and Cultivated Sunflowers

C. E. Rogers

"Sunflowers" is a botanical term which commonly refers to one or more of the 50 species and 19 subspecies of Helianthus that grows throughout the United States and into southern Canada and northern Mexico (Rogers et al. 1982). Several species of Helianthus may have been cultivated in prehistoric America, but only a few were truly domesticated (Heiser 1985). The prime example of a cultivated sunflower is the common annual sunflower, H. annuus L. The cultivated sunflower (Fig. 1) is the only native plant of the United States to become a major world-wide food source. Helianthus annuus (Fig. 2) is an extremely variable and widely distributed species (Heiser et al. 1969) that crosses with 19 species of Helianthus (Table 1). Habitats for wild Helianthus species vary from xeric (Fig. 3) to nearly constant wetness (Fig. 4). Hence, the wild species of Helianthus have become an invaluable genetic resource from which useful agronomic traits can be bred into cultivars (Laferriere 1986, Thompson et al. 1981). Examples of wild species value to agriculture include the incorporation of cytoplasmic male sterility into cultivars by crossing cultivated annuus with H. petiolaris Nuttall and the incorporation of cultivar resistance to major pathogens by crossing cultivated sunflower with H. tuberosus L. (Fick 1978, Zimmer and Hoes 1978). A sunflower germplasm depository has been established by the USDA, Agricultural Research Service to preserve endangered species for future agricultural prosperity (Rogers 1980a).

TABLE 1
General characteristics of wild sunflowers that are cross-compatible with the common annual sunflower, <u>H</u>. <u>annuus</u>, the cultivated species.*

| | General habitat | | Seed oil |
Sunflower species	Type	Avg. precipitation (cm/year)	content (%)
Annual			
<u>anomalus</u> Blake	sand	25-50	31.5-37.9
<u>argophyllus</u> T&G	sand	50-100	16.1-26.1
<u>bolanderi</u> Gray	valleys	25-150	23.8-27.2
<u>debilis</u> Nuttall	sand	64-140	19.0-34.9
<u>deserticola</u> Heiser	sand	<12-25	31.5-34.3
<u>neglectus</u> Heiser	sand	25-50	16.9-33.0
<u>niveus</u> Brandegee	sand	<12-50	26.8-40.2
<u>paradoxus</u> Heiser	wet soil	25+	25.2
<u>petiolaris</u> Nuttall	sand	25-127	20.6-37.4
<u>praecox</u> Engl. & Gray	sand	50-120	25.1-32.9
Hybrid cultivars	variable	45+	35.0-50.0
Perennial			
<u>angustifolius</u> L.	wet soil	90-175	25.2
<u>decapetalus</u> L.	shaded woodlands	60-140	21.3-24.5
<u>giganteus</u> L.	wet areas	50-140	20.5-32.5
<u>gracilentus</u> Gray	dry slopes	25-50	30.0

hirsutus Raf.	dry, open areas	63-140	15.2-25.6
maximiliani Schr.	prairies	25-127	20.8
rigidus (Cass.) Desf.	prairies	63-100	25.1-30.0
strumosus L.	variable	65-140	24.5
tuberosus L.	variable	50-140	20.5-31.4

* From Thompson et al. 1981; Rogers et al. 1982; Seiler 1985.

4

FIG. 1. Cultivated _Helianthus_ _annuus_ oilseed hybrid.

HISTORICAL

An abbreviated list of major historical events that shaped the current commercial sunflower industry in the United States occurs in Table 2. Archaeological finds suggest that sunflower (_H_. _annuus_ L.) was under cultivation by American Indians in Arizona and New Mexico as early as 3000 BC (Heiser 1976). by the time Europeans arrived in what was to become the United States, sunflower cultivation had extended throughout most of the area (Putt 1978). Indians

TABLE 2
Abbreviated history of cultivated sunflower.

Date	Activity and location
3,000 BC	Cultivated by American Indians for food, dye, and ceremony
1500–1700's	Introduced into Madrid; spread throughout Europe
1716	English patent for oil processing
1830's	Established as an oilseed crop in Russia
1875	Introduced into Canada by Russian immigrants
1900–1940	Emphasis in the United States on silage production
1914	First commercial seed production in the United States
1940's	Oilseed production began in Canada
1940's	Snack food industry developed in the United States
1960's	Beginning of commercial oilseed production in the U.S.
1972	First comnmercial hybrid varieties became available
1975	Expansion of oilseed acreage in Texas
1977	Major national expansion of oilseed acreage
1978–1979	Expansion of domestic oil markets; collection of wild species and establishment of germplasm depository in the United States
1980–1983	Construction of major seed crushing facilities in the United States
1980–1983	Concerted effort to use wild species germplasm for research on pest resistance, biochemistry, and interspecific crossing in the U.S.

FIG. 2. Wild (native) <u>Helianthus</u> <u>annuus</u> (common annual
sunflower) (Photo by G. J. Seiler).

used the sunflower nutmeat for food, seed wall and floral
pigments for dye, and plant residue for fuel. Spanish
explorers carried sunflower seed from the Southwest to
Madrid, Spain around 1500, and during the following 200
years sunflower spread throughout Europe as an ornamental
plant.

Sunflower became established as an oilseed crop in
Russia in the 1830's, and by 1900 it had mushroomed into
eastern Europe's primary source of vegetable oil (Putt
1978). Mennonite farmers immigrating from Russia in 1875

FIG. 3. Xeric habitat of <u>Helianthus anomalus</u> (across
middle of photograph) in sand dune (Photo by G. J.
Seiler).

introduced cultivated sunflower into Canada and began
planting it in home gardens. From that time until the
1940's, most commercial emphasis on sunflower production in
North America centered around its value as a silage crop
(Vinall 1922). In the 1940's, oilseed sunflower became a
viable crop in Canada, and confectionary sunflower became
an important snack food in the United States. The 1960's
saw major advances in the sunflower oilseed industry in the
United States. During that decade, seed yield and oil production were significantly increased, disease resistance
was incorporated into cultivars, and cytoplasmic male sterility was discovered (Heiser 1978). Hybrid varieties became commercially available to growers in 1972 (Fick 1978).
Major acreage expansions occurred in 1975, 1977, and 1979
in the United States and this country became the No. 2 sunflower producing country, behind the Soviet Union (Cobia
and Zimmer 1978). Large scale test marketing of sunflower

FIG. 4.   Swamp sunflower <u>Helianthus</u> <u>angustifolius</u>   (Photo
by G. J. Seiler).

oil products occurred in the U.S. during 1978-79, and major
new seed crushing facilities began construction during
1980.

MAJOR USES FOR SUNFLOWER

Today, about 15% of the "sunflower seed" produced in the
U.S. are the "edible" type, which is relatively low in oil

content (ca. 18-25%). Edible sunflower seed is used as confectionary in various products, eaten whole, and sold as bird seed (Cobia and Zimmer 1978). About 85% of the U.S. sunflower crop is the high oil type (averaging from 40 to > 50% oil content) hybrid varieties. Traditionally, from 85-95% of the oilseed sunflower seed produced in the U.S. has been exported to Western Europe. Sunflower oil is the premier vegetable oil crop in Europe, and Europeans will pay a premium price to obtain it. Primary use of sunflower oil in the U.S. has been for salad oils, cooking oils, and margarines (Cobia and Zimmer 1978). By-products obtained from processed sunflower oil include tocopheral, lecithin, waxes, phosphatides, distillates, filter clay, and glycerol. Defatted sunflower meal is used for animal food, and concentrates or isolates go into meat analogues and extenders, emulsifying agents, bakery products, beverages, and snack foods. Hulls from processed sunflower seed may find a use as fuel, feed, furfural, and particle boards (Dorrell 1978). In recent years, there has been considerable interest in using sunflower oil in vehicles as fuel. South Africa has converted a considerable proportion of its farm machinery to use sunflower oil as fuel (Bruwer et al. 1980). Another possible fuel source for sunflower lies in its high energy-value residue. The BTU's per pound of plant residue are higher for sunflower than for any other major crop but sugarcane (Oursbourn et al. 1978). The high BTU rating for sunflower residue is probably due to its unusually high rubber content in foliage and stems (Stipanovic et al. 1980, 1982).

Because sunflowers are native to the United States, their associated insect herbivores, and their entomophages, co-evolved in natural communities that often exist near cultivated fields of sunflower. The long association of wild sunflowers and insects in the U.S. has resulted in hundreds of species of insects that frequent the plants (Cockerell 1914, Robertson 1922, and Satterthwait 1948). Records of insects on sunflower are largely for the species that are associated with H. annuus. Insect faunas of the other 49 Helianthus species mostly are unknown. Fortunately, most species of insects associated with sunflowers are either innocuous or benefactors of the plants; relationships range from obligatory (McGregor 1976) to purely causal and non-essential. Many species of insects visit sunflowers to feed on their extrafloral nectar (Rogers 1985a), to obtain floral nectar and pollen for rearing their young (Parker 1981), to nurse ants and collect their excretions for food (Worth 1943), to seek out plant herbivores as hosts to either prey on or into which to lay their eggs (Rogers et al. 1972), to seek shelter and forage (Lynch and Garner 1980). Virtually every part of the sunflower plant is used by one or more species of insects (Genung and Green 1979, Phillips et al. 1973, and McBride et al. 1981). Species of insects that are recognized as pests of cultivated sunflower were summarized by Walker (1936), Beckham and Tippins (1976), Phillips et al. (1973), and Schulz (1978). Rajamohan (1976) and Rogers (1979a, 1980b) compiled bibliographies of insect pests of sunflower, and noted the natural enemies of the pests.

CROP PROTECTION

Crop protection has been a constant concern of sunflower agriculturists in the United States (Schulz 1978), as well as in Africa (Khaemba and Mutinga 1982), Asia (Ayyanna et al. 1978), Australia (Broadley 1978), Europe (Stoenescu et al. 1974), and South America (Orfila 1964). However, pest problems on sunflower have been more acute in the United States than elsewhere due to the natural co-evolutionary associations of sunflower and its pests. Sunflower resistance to pathogens has been successfully used to control serious diseases in cultivars on every continent (Zimmer and Hoes 1978). Europeans also use host resistance to protect sunflower crops from the European sunflower moth, _Homoeosoma nebulella_ Hubner (Shapiro 1975). While host resistance has been a useful strategy for reducing seed losses caused by insects in Europe, American sunflower breeders and entomologists have relied on chemical pesticides for controlling insect pests (Archer et al. 1983, Carlson 1971, Teetes and Randolph 1968, Charlet et al. 1985). Only in recent years has there been a real interest in exploiting sunflower's natural defenses for control of economically important insect pests in the United States (Rogers 1981a, Rogers and Thompson 1978a, Gershenzon et al. 1985, and Rogers et al. 1987). This paper will discuss our limited knowledge of entomology of indigenous sunflowers and how such knowledge may be used to enhance management of pest species in cultivated sunflower.

LEPIDOPTERA PESTS OF SUNFLOWER

Several species of Lepidoptera use wild sunflowers as a host and attack cultivated sunflower (Table 3). Historically, the sunflower moth, _Homeosoma electellum_ (Hulst) has been, and continues to be, the most damaging insect pest of cultivated sunflower in the U.S. Intensifying the economic impact of the sunflower moth on cultivated sunflower is the predisposition of damaged capitula to become infected with _Rhizopus_ spores (Klisiewicz 1979, Rogers et al. 1978d), resulting in reduced seed yield and rancid oil due to systemic infection by the fungus (Thompson et al. 1980). The sunflower moth is distributed throughout the U.S. (Heinrich 1921), and it has become a pest wherever the crop has become established (Lynch and Garner 1980, Rogers 1978, Johnson and Beard 1977, and DePew 1983).

Numerous species of parasites and predators attack the larva of the sunflower moth, but they have been ineffective in maintaining subeconomic populations of larvae in cropping systems (Rogers 1980a, Teetes and Randolph 1969a, and Beregovoy 1985a). The ineffectiveness of natural enemies to control larval populations of the sunflower may be due to the moth's habit of making long-range migrations into distant environments (Rogers et al. (1986a). Although cultural control of _H. electellum_ has been suggested (Teetes and Randolph 1969b, 1971), the use of chemical pesticides remains the control strategy of choice by necessity (Carlson 1971, Teetes and Randolph 1971, Chandler and Heilman 1982, Archer et al. 1983, Rogers et al. 1984a). Commercially available sex pheromone formulations now provide an economical and efficient means for monitoring adult populations in commercial sunflower (Underhill et al. 1979, Rogers and Underhill 1983). Although current technology provides good control of the sunflower moth, a better understanding of the natural host/insect relationship may provide more economical and persistent management possibilities.

The sunflower moth uses several species of the Compositae, including a few _Helianthus_ spp. as natural hosts (Drake and Harris 1926, Teetes and Randolph 1969c, Beregovoy 1985b). Young larvae of _H. electellum_ feed primarily on floral structures and pollen (Rogers 1978). While older larvae may continue to feed on pollen and floral structures, they feed more on immature, pliable achenes (seeds) and tissue of the capitulum (sunflower head). Even though larvae of _H. electellum_ are commonly found in the capitula of wild sunflower heads, they do only limited damage to the

TABLE 3
Lepidoptera pests of cultivated sunflower.

Species	Common name	Plant part attacked
Chlosyne lacinia Geyer (Gelechiidae)	Checkerspot butterfly	foliage
Cochylis hospes Walsingham (Cochylidae)	Banded sunflower moth	seed
Eucosma womonana Kearfott (Tortricidae)	none	root, lower stalk
Euxoa messoria Harris (Noctuidae)	Darksided cutworm	seedling
Euxoa ochrogaster Guenne (Noctuidae)	Redbacked cutworm	seedling
Homoeosoma electellum Hulst (Pyralidae)	Sunflower moth	seed, receptacle floret, upper stalk
Isophrictis similiella Chambers (Gelechiidae)	none	lower stalk
Stibadium spumosum Grote (Noctuidae)	none	seed, receptacle
Suleima helianthana Riley (Tortricidae)	Sunflower budmoth	receptacle, stalk
Vanessa cardui (L.) (Nymphalidae)	Painted lady	foliage

achenes (Rossiter et al. 1986). The pericarp in achenes of wild sunflowers matures and hardens fairly rapidly, and this rapid hardening protects the achene from damage by younger larvae. Maturation of achenes in individual capitula in wild sunflowers from flowering to the beginning of senescence approximates the developmental period of H. electellum larvae. The relatively rapid hardening of the pericarp in wild sunflowers deters larval feeding, thereby forcing larvae to remain partially exposed while they feed on floral structures; exposure also results in a high degree of larval parasitism and predation on wild species of hosts (Beregovoy 1985a).

Premature hardening of sunflower achenes is caused by the deposition of a phytomelanin layer between the epidermis and hypodermis of the pericarp (Rogers and Kreitner 1983). The phytomelanin becomes extremely dense after its deposition in the pericarp, making mature achenes with this characteristic relatively more resistant to mechanical puncture (Stafford et al. 1984). In some breeding lines of cultivated sunflower, the phytomelanin begins developing within 3 days after fertilization of the achene (Rogers and Kreitner 1983). All species of Helianthus have phytomelanin-rich pericarp, although the characteristic appears to not be present in all populations of a species (Seiler et al. 1984). Bioassays indicate that a chemical factor associated with the phytomelanin of immature achenes provides the achenes with this characteristic some additional protection against young larvae (Johnson and Beard 1977, Rogers and Kreitner 1983). Recent chromatographic analyses of immature achenes showed that pericarp having the phytomelanin possesses a chemical(s) that does not occur in pericarp lacking the phytomelanin (Severson and Rogers, unpublished data). Three sunflower moth-resistant germplasm lines released in 1984 have phytomelanin in the pericarp (Rogers et al. 1984b).

Chemical extracts from H. annuus florets contain diterpenoid acids that produce a chronic lengthening of larval stadia but not a high mortality (Waiss et al. 1977 Rossiter et al. 1986). More recently, a sesquiterpene lactone extracted from glandular trichomes at the anther apex of H. maximiliani Schrader causes a high acute mortality of H. electellum larvae in laboratory bioassays (Gershenzon et al. 1985, Rossiter et al. 1986, and Rogers et al. 1987). The chemistry of Helianthus species is just now becoming sufficiently well known to permit us to take advantage of this knowledge in the management of the sunflower moth and other pestiferous insect species on cultivated sunflower.

14

Other lepidopterous pests of cultivated sunflower are either of local concern, or only periodically become economically important. The banded sunflower moth, _Cochylis hospes_, has been an acute problem in sunflower in the northern Plains since the early 1980's. Feeding habits of _C. hospes_ larvae and their damage to sunflower resemble the behavior and damage inflicted by larvae of _H. electellum_ (Westdal 1949 and McBride et al. 1985). Although little is known about the ecology of _C. hospes_, I collected larvae of this species in capitula of _H. annuus_ and _H. argophyllus_ Torrey & Gray in Texas (unpublished data). The only option for control of _C. hospes_ in cultivated sunflower at this time is the use of chemical pesticides (Charlet and Busacca 1986). A recently described sex pheromone for _C. hospes_ has become a very useful survey tool for this pest (Underhill et al. 1986).

Cholsyne lacinia (checkerspot) larvae gregariously defoliate adjacent sunflower plants in restricted areas in the southern Plains (Stamp 1977). The checkerspot has not become an economic threat to cultivated sunflower. Its wide host range includes _H. annuus_, _H. debilis_ ssp. _cucumerifolius_ (Torrey & Gray), _H. maximiliani_, _H. ciliaris_ DC, _H. tuberosus_, _H. petiolaris_ ssp _petiolaris_ Nuttall, _H. argophyllus_, and several closely related species of compositae (Neck 1973, 1977).

Stibadium spumosum is a widely distributed noctuid whose large larvae hollow-out large areas in the inflorescence as they consume immature sunflower seed (Rogers et al. 1986b). This species attacks wild _H. annuus_ throughout the southeastern 2/3 of the United States (Comstock 1946). Often, one may find a high percentage of the _H. annuus_ capitula of a population infested with _S. spumosum_ larvae. Although larvae commonly occur in cultivated sunflower in the southern Plains, this species is not considered an economic threat at this time.

The larva of _Eucosma womonana_ is a root borer that has caused extensive damage in sunflower research plots on the Texas High Plains (Rogers et al. 1979b). _Eucosma womonana_ larvae have been collected from the roots of _H. annuus_, _H. divaricatus_ L., and _H. tuberosus_ (Rogers 1985b). Two related species, _E. sombreanna_ Kearfott and _E. wandana_ Kearfott were reared from commercial _H. tuberosus_ tubers in Minnesota in 1983 (Niehaus, personal communication). Although not much is known about its bionomics, _E. womonana_ appears to be univoltine, and adults are easily captured in traps baited with its recently described sex pheromone (Underhill et al. 1987). Like _E. womonana_, adults of

Isophrictis *similiella* are seldom seen, except when removed from sex pheromone traps placed near the soil surface (Underhill et al. 1987). Females of I. *similiella* lay their eggs on the sunflower stalk near the ground, and larvae become stalk borers. Though not an economic pest, I. *similiella* larvae are common in cultivated annuus stalks in the Plains states, and have been collected from wild H. annuus (Charlet 1983c).

Vanessa *cardui* (painted lady) is a migratory species whose larva is capable of defoliating entire sunflower fields with little advance notice of its presence (Schulz 1978). In addition to sunflower, several other crops are subject to defoliation by larvae of the painted lady (Poston et al. 1977). Large populations of V. *cardui* larvae may appear anywhere in the U.S. during the summer, but adults migrate to Mexico to spend the winter (Williams 1970). Were it not for its high mortality due to parasitism (Schrader 1921, Rogers 1980b), the larva of this species would be much more destructive to crops. The larva is easily killed by pesticides that are applied for its control or for the control of H. *electellum* larvae. Wild H. annuus is among its most common natural host.

Cutworms, primarily larvae of *Euxoa* *messoria* and E. *ochrogaster*, are perennial pests of seedling sunflower in the northern Plains (Schulz 1978). Larvae often cause considerable loss of plants in isolated areas, and can be controlled by pesticidal applications. Nothing is known of the bionomics of these species on their indigenous hosts.

Suleima *helianthana* (sunflower bud moth) is a pest of cultivated sunflower throughout the Plains States (Ehart 1974, Rogers 1979b). In southern latitudes, first generation larvae occur most frequently in the growing terminal of the plant. Second generation larvae more commonly attack the plant capitulum, often causing considerable seed loss on infested plants. Although activity of larvae in the plant often causes dramatic injury symptoms, economic losses seldom occur (Rogers 1979b). Insecticides appear to be ineffective in reducing larval populations in cultivated sunflower and are of questionable value in their control. In addition to H. annuus, S. *helianthana* has been collected from 16 other *Helianthus* species (Charlet 1983c and Rogers 1979b).

COLEOPTERA PESTS OF CULTIVATED SUNFLOWER

The curculionids, as a group, have been the most economically damaging insect pests of cultivated sunflower in the Great Plains states. Weevil larvae attack every part of the plant and often inflict severe tissue destruction and yield loss before their presence is evident (Table 4).

Cylindrocopturus adspersus has been particularly troublesome, having emerged as a pest of cultivated sunflower in eastern Colorado in 1920 (Newton 1921). Wherever sunflower has become established as a crop west of the Mississippi River, C. adspersus has become a serious pest (Phillips et al. 1973, Casals-Bustos 1976, and Rogers et al. 1983). Females of C. adspersus lay eggs just beneath the epidermis on the lower stalk of the plant (Charlet 1983a), and subsequent development occurs as larvae burrow through the stalk (Rogers et al. 1983). Larvae eventually find their way to the stalk base and upper root crown where they spend the winter (Charlet 1983b, Rogers and Serda 1982). Cylindrocopturus adspersus has a single generation a year, and cultural strategies such as fall-winter-spring tillage and altering planting dates serve a minor role in reducing plant loss (Oseto et al. 1982, Rogers and Jones 1979, Rogers et al. 1983). The use of insecticides also sometimes appears of benefit in protecting plants against destruction by larvae, although the economic benefits from their use may be questioned (Charlet and Oseto 1983, Rogers et al. 1983). Yield loss in sunflower is caused by C. adspersus larval tunneling that weakens the stalks and interferes with the plant's water uptake and transport ability (Rogers et al. 1983, Rogers and Jones 1979), and by the insect's mechanical transmission of the fungus Macrophomina phaseolina (Tassi) Goid, the causative agent of charcoal rot and premature ripening (Yang et al. 1983, Yang and Owen 1982).

Several species of Compositae are used as natural hosts by C.adspersus (Hilgendorf and Goeden 1981), including H. annuus, H. petiolaris, and H. maximiliani (Charlet 1983c). Immature stages of C. adspersus are attacked by several parasites and predators (Charlet 1983d, and Charlet and Balsbaugh 1984), which appear to have no impact on host populations. Many Helianthus species (ca. 25) have significant levels of resistance to C. adspersus female oviposition and/or larval development (Rogers and Seiler 1985). Hence, the use of host resistance through germplasm derived from wild Helianthus species holds considerable promise as an enduring, cheap management strategy for

TABLE 4
Coleoptera pests of cultivated sunflower.

Species	Common name	Plant part attacked
Apion occidentale Fall (Curculionidae)	Black sunflower stem weevil	stalk, petiole
Ataxia hubbardi Fisher (Cerambycidae)	None	stalk, petiole
Baris strenua (LeConte) (Curculionidae)	Sunflower root weevil	root crown
Bathynus gibbosus (DeGeer) (Scarabaeidae)	Carrot beetle	root
Cylindrocopturus adspersus (LeConte); (Curculionidae)	Spotted sunflower stem weevil	stalk, root
Dectes texanus LeConte (Cerambycidae)	Soybean girdler	stalk
Haplorhynchites aeneus (Boheman); (Curculionidae)	Sunflower clipping weevil	peduncle
Mecas inornata Say (Cerambycidae)	Sunflower girdler	stalk
Smicronyx fulvus LeConte (Curculionidae)	Red sunflower seed weevil	seed
Smicronyx sordidus LeConte (Curculionidae).	Gray sunflower seed weevil	seed
Zygogramma exclamationis (Fab.); (Chrysomelidae)	Sunflower beetle	foliage

18

<u>C</u>. <u>adspersus</u> in cultivated sunflower.

Four species of weevils that are common but of minor economic importance on cultivated sunflower are <u>B</u>. <u>strenua</u>, <u>A</u>. <u>occidentale</u>, <u>Rhodobaenus</u> <u>tredecempunctatus</u> Illiger (cocklebur weevil), and <u>H</u>. <u>aeneus</u> (head clipper) (Table 4). Adults of <u>B</u>. <u>strenua</u> feed on foliage but are of no economic importance. However, <u>B</u>. <u>strenua</u> are common borers in the cortex of the root crowns of cultivated sunflower, particularly in the northern Plains (Casals-Bustos 1976). Eggs of <u>B</u>. <u>strenua</u> are placed·just beneath the epidermis on the root crown, where larvae subsequently burrow as they feed. <u>Baris</u> includes many species of weevils that frequent Compositae as natural hosts, including <u>H</u>. <u>annuus</u>, <u>H</u>. <u>petiolaris</u>, <u>H</u>. <u>maximiliani</u>, and <u>H</u>. <u>grosseserratus</u> Martens (Casals-Bustos 1976). Natural parasitism of immature <u>B</u>. <u>strenua</u> appears to be less than 2%. The black sunflower stem weevil (<u>A</u>. <u>occidentale</u>) is very common on cultivated sunflower, but damage caused by direct feeding of its larva is insignificant (Gaudet and Schulz 1981). The major importance of <u>A</u>. <u>occidentale</u> lies in its ability to transmit fungi, <u>Phoma</u> <u>oleracea</u> var. <u>helianthi-tuberosi</u> Sacc. and <u>P</u>. <u>macdonaldi</u> Boerema, the causative agent of phoma black stem to cultivated sunflower (Gaudet and Schulz 1981, 1984). Symptoms of <u>Phoma</u> infection include blackened leaves, petioles, and nodes of infected plants. <u>Phoma</u> may contribute to the premature ripening syndrome of cultivated sunflower. Little is known about the bionomics of <u>A</u>. <u>occidentale</u> apart from its existence on cultivated sunflower. The cockle-bur weevil (<u>R</u>. <u>tredecempunctatus</u>), a large, brick-red insect with black splotches on its elytra and prothorax, is distributed throughout the United States (Vaurie 1981). Although the cockle-bur weevil is usually not considered as an economic threat to cultivated sunflower, its larvae destroyed entire fields of sunflower in western Tennessee in 1981 (personal observation). The infested fields also were heavily populated by the giant ragweed, which is a favored host of <u>R</u>. <u>tredecempunctatus</u>. This insect commonly infests several species of the Compositae, including <u>Ambrosia</u>, <u>Xanthium</u>, <u>Oenothera</u>, and <u>Helianthus</u> genera (Weiss and Lott 1923, Genung and Green 1983). Adults of the cockle-bur weevil oviposit in the peduncle of the plant, and larvae subsequently burrow to the roots to overwinter (Chittenden 1922). Management of nearby weed hosts may be helpful in decreasing infestation of cultivated sunflower by <u>R</u>. <u>tredecempunctatus</u>. The sunflower headclipping weevil, <u>H</u>. <u>aeneus</u>, is a large black weevil that is widely distributed in the United States and commonly seen in cultivated sunflower. Adults begin visiting sunflower after pollen becomes abundant on the inflorescence. After feeding on pollen (and possibly nectar), the females move to just below the capitulum and girdle the peduncle, after which they return to the inflorescence to lay their eggs (Hamilton 1973). The girdled capitulum hangs on the peduncle for a few days and drops to the ground, where larval development occurs in the decaying tissue of the capitulum. Haplorhynchites aeneus uses several species of Silphium as hosts (Hamilton 1974), and has been reared from H. annuus, H. divaricatus L., H. grosseserratus, and H. microcephalus Torrey & Gray (Hamilton 1973, 1974). As with all weevil species attacking cultivated sunflower, H. aeneus is univoltine.

The sunflower seed weevils, Smicronyx fulvus and S. sordidus, are serious pests of cultivated sunflower and have been responsible for the termination of sunflower production in some areas (Satterthwait 1946). Both species are widely distributed in the United States, although their relative abundance changes with geography. S. fulvus is more abundant from the northern Plains to the High Plains of Texas, while S. sordidus becomes more abundant in the Texas Rolling Plains and southward (Heilman et al. 1983, Oseto and Braness 1979, and personal observations). Females of both species lay eggs in immature seeds, where a single larva per seed develops (Oseto and Braness 1979). Larvae hollow out the achene and drop to the ground to spend the winter and pupate (Oseto and Charlet 1981, Heilman et al. 1983). Infestations of sunflower fields by Smicronyx usually remain higher along field margins than further into fields (Charlet and Oseto 1982). Several species of parasites attack larvae of Smicronyx (Bigger 1930, Satterthwait 1946, Oseto and Braness 1979). Nevertheless, weevil populations often become economically threatening in cultivated sunflower and require control by pesticides (Oseto and Braness 1980, and Gednalske and Walgenbach 1984). Natural hosts of seed weevils seem to be primarily Helianthus species, including annuus, maximiliani, tuberosus, hirsutus, and deserticola (Satterthwait 1946, Oseto and Braness 1979, and personal observation).

Numerous species of Cerambycidae frequent sunflower, but only Ataxia hubbardi, Dectes texanus, and Mecas inornata have persistently attacked cultivated sunflower in the United States. These species are univoltine, except that D. texanus may be bivoltine in southern latitudes (Rogers 1977a). Dectes texanus is the most damaging cerambycid pest of cultivated sunflower; it also is a serious pest of soybean in the Southeast (Hatchett et al. 1975). Females may girdle the stalk (M. inornata), leaf petioles (A. hubbardi), or scar the stalks (D. texanus) prior to placing an egg just beneath the epidermis (Rogers 1977a). Developing larvae subsequently burrow through the pith of the stalk as they make their way to the root. In the fall, the mature larvae girdle the base of the stalk from the inside to the epidermis. They then pack the gallery of the root crown below the girdled plant and spend the winter in the sunflower root. There currently is no effective management strategy for escaping sunflower injury by cerambycids. Soil tillage, delaying the planting date, and pesticides may be of questionable benefit in reducing losses caused by these insects (Rogers 1985c). Wild species of Helianthus are among the natural hosts for cerambycids. Mecas inornata has been collected from H. annuus, H. maximiliani, H. tuberosus, H. mollis Lambert, and H. salicifolius (Rogers 1977a). Ataxia hubbardi and D. texanus occur less frequently on Helianthus species, although both D. texanus and D. sayi Dillon and Dillon have been collected from H. annuus (Charlet 1983a, Genung and Green 1983), and A. hubbardi from H. annuus (Muma et al. 1950) and Helianthus species (Fisher 1924).

Bothynus gibbosus (carrot beetle) is a widely distributed pest of several crops in the United States (Chittenden 1902). As its common names implies, the carrot beetle prefers carrots and other cultivated root crops, and several crops that have large tap roots (Hayes 1917). Unlike most pestiferous scarabs, the adult is the damaging stage of B. gibbosus (Bottrell et al. 1973, Rogers 1974). The nocturnal adults can invade a field of sunflower, burrow into the soil, and denude the roots before their presence is evident (Rogers 1974). As few as two beetles per plant can destroy enough root tissue to severely interfere with plant uptake of soil water and nutrients (Brigham et al. 1974). There are no effective control strategies for protecting crops from the carrot beetle. Even insecticides applied to the base of plants fail to prevent crop loss caused by adult feeding on roots (Rogers and Howell 1973, Brigham et al. 1974). Adult B. gibbosus prefer sandy soils and attack the root of plants found in these habitats. Helianthus petiolaris is among the preferred natural hosts of B. gibbosus (Rogers 1974). In nursery plantings of Helianthus at Munday, TX, the roots of H. annuus, H. argophyllus, and H. maximiliani were severely damaged by the carrot beetle, while the roots of H. hirsutus, H. tuberosus, and H. mollis Lambert showed no evidence of injury (Rogers and Howell

1973). Laboratory and greenhouse studies showed that about one-half the wild Helianthus species are resistant to carrot beetle injury (Rogers et al. 1980, Rogers and Thompson 1978b). Carrot beetles feeding on the roots of H. arizonensis Jackson, H. atrorubens, H. occidentalis ssp. plantagineous Torrey & Gray, and H. porteri Blake suffered a high acute mortality in no choice feeding tests (Rogers et al. 1980).

The sunflower beetle, Zygogramma exclamationis, has been recognized as an important defoliator of sunflower since the late 1800's Chittenden 1988). Although this species has been considered economically important to sunflower production in the northern Plains (Criddle 1922), it is not a serious pest in the southern Plains (Rogers 1977b). The sunflower beetle defoliates primarily sunflower in both larval and adult stages (Rogers 1977b). Adults are active during the day, but larvae are primarily nocturnal feeders. In some years, seedling plants in the northern Plains are subject to extensive defoliation by Z. exclamationis and require the application of insecticides for their protection against denudation (Schulz 1978). Although Z. exclamationis is considered to feed primarily on Helianthus species (Brisley 1925 and Criddle 1922), I observed it feeding on wooly leaf bursage, Franseria tomentosa Gray, in the Texas Panhandle (Rogers 1977b). In laboratory studies, about one-half of the wild sunflower species exhibited resistance to feeding and/or reproduction by Z. exclamationis (Rogers and Thompson 1978c, 1980). Antibiosis against both larvae and adult beetles was strongly expressed, particularly by the perennial Helianthus species. It appears that the incorporation of germplasm from the Helianthus species in sunflower has real merit as a management strategy for Z. exclamationis and other coleopterous pests.

DIPTERA PESTS OF SUNFLOWER

Three species of tephritids commonly attack cultivated sunflower (Table 5) but their damage is mostly subeconomic (Kamali 1973). Strauzia longipennis (sunflower maggot) females lay their eggs just below the epidermis of stalks on young plants (Westdal and Barrett 1960), where subsequent larval tunneling causes mechanical injury to the pith and predisposes the plant to Sclerotinia sclerotiorium (Lib.) de Bary (Westdal and Barrett 1962). Stalk damage is particularly devastating to sunflower grown for silage (Caesar

TABLE 5
Diptera pests of cultivated sunflower.

Species	Common name	Plant part attacked
Contarinia schulzi Gagne (Cecidomyiidae)	Sunflower midge	receptacle, tissue
Gymnocarena diffusa Snow (Tephritidae)	None	floral tissue
Neolasioptera helianthi Felt (Cecidomyiidae)	Sunflower seed midge	seeds
Neotephritis finalis (Loew) (Tephritidae)	None	seeds
Strauzia longipennis (Wied.) (Tephritidae)	Sunflower maggot	stalks

1924). The sunflower maggot is found throughout the United States and southern Canada (Westdal and Barrett 1960), and occurs as several biotypes (Kamali 1973). In some years, nearly all the sunflower plants in northern latitudes may be infested with larvae of S. longipennis (Kamali 1973, Westdal and Barrett 1962). This univoltine fly has been collected from wild H. annuus, H. maximiliani, and H. tuberosus (Westdal and Barrett 1960), and can be controlled by insecticidal applications should the need arise (Allen et al. 1954).

Gymnocarena diffusa is a univoltine tephritid whose larvae burrow and feed on the tissue of the sunflower capitulum for up to 30 days before pupating (Kamali 1973). Alternate hosts for G. diffusa include H. annuus and H. maximiliani (Kamali and Schulz 1974). Neotephritis finalis is one of the most common tephritids in the United States (Foote 1960), is widely distributed, and probably the most damaging tephritid attacking sunflower (Kamali 1973). Several dozen adults may be reared from a single capitulum in southern latitudes (Beckham and Tippins 1972). Females of N. finalis oviposit among corollas of the inflorescence, where larvae burrow into immature seed to develop and pupate. This polyphagous species attacks several species of composites, including Helianthus species, and is multivoltine in southern latitudes.

Two additional species of tephritids of no economic consequence to commercial sunflower production are common in wild sunflower of southern California. Traupanea bisetosa (Coquillet) larvae develop in seeds of cultivated sunflower, wild H. annuus, and and H. ciliaris (Decandolle) (Cavender and Goeden 1982). Paracantha cultaris (Coquillet) attack unopened flower heads of H. annuus, resulting in deformed and stunted capitula, and as high as 50% seed destruction (Cavender and Goeden 1984).

Several species of cecidomyiid midges attack cultivated sunflower and about 35 species of Helianthus (Rogers et al 1979a). However, only two species are considered to be of economic importance. Contarinia schulzi (sunflower midge) is a serious pest of cultivated sunflower along the Red River Valley of North Dakota and Minnesota (Kopp 1983). Although C. schulzi is economically important only in northern Plains sunflower, I have collected it from H. annuus in Kansas and Texas, and from H. petiolaris in Kansas (Rogers 1977c, 1982). It also has been collected from H. annuus, H. petiolaris, and H. maximiliani in North Dakota (Schulz 1973). Larvae of C. schulzi feed on vegetative tissue of the inflorescence, where the damage becomes

24

manifested by necrotic tissue and knarled capitula (Schulz 1973). Infestations by C. schulzi may severely reduce seed production in commercial sunflower (Fick and Auwater 1981). Although chemicals frequently have been applied for control of C. schulzi, their effectiveness in economic returns is questionable (Busacca 1983). Altering planting date appears to reduce midge infestations in some instances (Lofgren 1983), and the use of midge-resistant hybrids is receiving considerable research emphasis (Fick and Auwater 1981).

The sunflower seed midge, Neolasioptera helianthi, is the most widely distributed and common midge infesting commercial sunflower (Rogers et al. 1979a). Several hundred adults have been reared from a single capitulum from cultivated sunflower (Rogers 1977c). Females of the sunflower seed midge lay their eggs among disk flowers of the inflorescence, where larvae burrow into immature seeds for development and pupation (Kreitner and Rogers 1981). This midge has been reared from the capitula of 31 Helianthus species (Rogers et al. 1979a). There are no proven control strategies for N. helianthi, although this and other midges from sunflower are heavily parasitized (Rogers 1977c).

MISCELLANEOUS INSECTS

Several other insect species attack cultivated sunflower but are of no general economic concern. Aphids and leafhoppers are sometimes abundant (Rogers et al. 1978a, Rogers 1981b). Fortunately, species of these insects (as well as others, e.g.midges) are usually restricted to either annual or perennial Helianthus species (Rogers et al. 1979a, 1978d). Therefore, knowing the host range of insects on wild sunflower should be a key to understanding their potential pest status on cultivated sunflower. Laboratory and greenhouse studies showed that several species of Helianthus that are cross-compatible with cultivated sunflower are resistant to attack from Masonaphis masoni (Knowlton) (Aphididae) and the western potato leafhopper, Empoasca abrupta DeLong (Cicadellidae) (Rogers 1981b, Rogers and Thompson 1978d, 1979).

Several insect pests of cultivated sunflower are controlled by the use of chemical pesticides. However, the economic advantages of this control tactic have been difficult to document. Other management strategies, e.g., cultural, selective planting date, biological control, and host resistance also are effective in reducing populations

FIG. 5. Wild <u>H. annuus</u> x cultivated sunflower outcross
that survived defoliation in a field that was
destroyed by larvae of <u>V. carduii</u>.

of select pest species (Fig. 5). A dramatic example of the potential usefulness of wild Helianthus species for enhancing pest resistance in cultivated sunflower is seen in Figure 5. Larvae of the painted lady butterfly, Vanessa cardui (L.) (Lepidoptera: Nymphalidae) denuded late-planted fields of cultivated sunflower on the Texas High Plains in the fall of 1978 and 1979. Among the cultivars were a few wild-type hybrid outcrosses with wild annuus that were avoided by V. cardui (Fig. 5). Chemical analyses of the plants showed that the foliage of the cultivar contained the diterpene (-) kaur-16-en-19-oic acid at 0.15% concentration. Foliage from the outcrosses contained 0.38% (-)-kaur-16-en-19-oic acid, 0.03% trachylobanic acid, and 0.05% angeloylgrandifloric acid. Host resistance via incorporation of germplasm from wild Helianthus offers a tremendous potential for long-lasting, economical management of several insect pests of cultivated sunflower. For the short term it appears that if entomologists would become familiar with the bionomics of pests on their native Helianthus hosts, much could be learned about ecologically sound management of insects on cultivated sunflower.

REFERENCES CITED

Allen, W. R., P. H. Westdal, C. F. Barrett, and W. L. Askew. 1954. Control of the sunflower maggot, Strauzia longipennis (Wied.) (Diptera: Trypetidae) with Demeton. Ann. Rept. Entomol. Soc. Ont. 85: 53-56.

Archer, T. L., C. S. Manthe, C. E. Rogers, and E. D. Bynum, Jr. 1983. Evaluation of several insecticides for control of the sunflower moth. The Southwest. Entomol. 8: 54-56.

Ayyanna, T., G. V. Subbartnam, and E. Dharmaraju. 1978. Pest complex on sunflower, Helianthus annuus L., in Andhra Pradesh. Ind. J. Entomol. 40: 353-357.

Beckham, C. M., and H. H. Tippins. 1972. Observations on sunflower insects. J. Econ. Entomol. 65: 865-866.

Beregovoy, V. 1985a. Parasitism of the sunflower moth, Homoeosoma electellum (Hulst) (Lepidoptera: Pyralidae) in the central United States. J. Kans. Entomol. Soc. 58: 732-736.

Beregovoy, V. 1985b. Appearance of first generation larvae of the sunflower moth, Homoeosoma electellum (Hulst) (Lepidoptera: Pyralidae), in the central United States. J. Kans. Entomol. Soc. 58: 739-742.

Bigger, J. H. 1930. A parasite of the sunflower weevil. J. Econ. Entomol. 23: 287.

Bottrell, D. G., R. D. Brigham, and L. B. Jordan. 1973. Carrot beetle: pest status and bionomics on cultivated sunflower. J. Econ. Entomol. 66: 86-90.

Brigham, R. D., C. A. Schaefer, and G. L. Teetes. 1974. Latest developments in biology and control of the carrot beetle, Bothynus gibbosus De Geer, a pest of sunflower in parts of the U.S.A. Proc. 6th Internat'l. Sunflower Assoc. 670-687. Bucharest, Romania.

Brisley, H. R. 1925. Zygogramma exclamationis. In. Notes on the Chrysomelidae of Arizona. Trans. Amer. Entomol. Soc. 874: 167-182.

Broadley, R. H. 1978. Insect pests of sunflower. Queensland Agric. J. 104: 307-314.

Bruwer, J. J., B. van D. Bashoff, F. J. C. Hugo, L. M. du Plessis, J. Fuls, C. Hawkins, A. N. van der Walt, and A. Engelbrecht. 1980. Sunflower seed oil as an extender for diesel fuel in agricultural tractors. Proc. 1980 Symp. So. Africa. Inst. Agric. Engineers. Silverton, 7 pages.

Busacca, J. D. 1983. Effectiveness of chemical insecticides in controlling the sunflower midge. Proc. Sunflower Res. Workshop. page 20.   Nat'l. Sunflower Assoc., Fargo, ND.

Caesar, L. 1924. The sunflower maggot -- a pest of sunflowers. Farmer's Advocate 59: 990-991.

Carlson, E. C. 1971. New insecticides to control sunflower moth. J. Econ. Entomol. 64: 208-210.

Casals-Bustos, P. 1976. Bionomics of Cylindrocopturus adspersus LeC. and Baris strenua (LeC.) (Coleoptera: Curculionidae). Thesis. 108 pages. North Dakota State Univ., Fargo.

Cavender, G. L., and R. D. Goeden. 1982. Life history of Traupanea bisetosa (Diptera: Tephritidae) on wild sunflower in southern California. Ann. Entomol. Soc. Amer. 75: 400-406.

Cavender, G. L., and R. D. Goeden. 1984. Life history of Paracantha cultaris (Coquillett) on wild sunflower, Helianthus annuus L. spp. lenticularis (Douglas) Cockerell, in southern California (Diptera: Tephritidae). Pan-Pacific Entomol. 60: 213- 218.

Chandler, L. D., and M. D. Heilman. 1982. Efficacy and phytotoxicity evaluation of microbial insecticides for control of the sunflower moth in south Texas. PR-4056: 5 pages. Texas Agric. Exp. Sta., Texas A&M Univ., College Station.

Charlet, L. D. 1983a. Ovipositional behavior and site selection by a sunflower stem weevil, Cylindrocopturus adspersus (Coleoptera: Curculionidae). Environ. Entomol. 12: 868-870.

Charlet, L. D. 1983b. Distribution and abundance of a stem weevil, Cylindrocopturus adspersus (Coleoptera: Curculionidae), in cultivated sunflower in the northern Plains. Environ. Entomol. 12: 1526-1528.

Charlet, L. D. 1983c. Insect fauna of native sunflower species in western North Dakota. Environ. Entomol. 12: 1286-1288.

Charlet, L. D. 1983d. Parasitoids of a stem weevil, Cylindrocopturus adspersus (Coleoptera: Curculionidae), in sunflower: incidence and parasitism rates in the northern Great Plains. Environ. Entomol. 12: 888-890.

Charlet, L. D., and E. U. Balsbaugh, Jr. 1984. Anaphes conotracheli (Hymenoptera: Mymaridae), an egg parasitoid of Cylindrocopturus adspersus (Coleoptera: Curculionidae). J. Kans. Entomol. Soc. 57: 526-528.

Charlet, L. D., and J. D. Busacca. 1986. Insecticidal control of banded sunflower moth, Cochylis hospes (Lepidoptera: Cochylidae), larvae at different sunflower growth stages and dates of planting in North Dakota. J. Econ. Entomol. 79: 648-650.

Charlet, L. D., and C. Y. Oseto. 1983. Toxicity of insecticides on a stem weevil, Cylindrocopturus adspersus (Coleoptera: Curculionidae), and its parasitoids in sunflower. Environ. Entomol. 12: 959-960.

Charlet, L. D., and C. Y. Oseto. 1982. Spatial pattern of Smicronyx fulvus (Coleoptera: Curculionidae) in cultivated sunflower, based on damage assessment. J. Kans. Entomol. Soc. 55: 351-353.

Charlet, L. D., C. Y. Oseto, and T. J. Gulya. 1985. Application of systemic insecticides at planting: effects on sunflower stem weevil (Coleoptera: Curculionidae) larvae numbers, plant lodging, and seed yield in North Dakota. J. Econ. Entomol. 78: 1347-1349.

Chittenden, F. H. 1898. A leaf beetle injurious to cultivated sunflower. USDA Bur. Entomol. Bull. 18:96.

Chittenden, F. H. 1902. Some insects injurious to vegetables. USDA Bur. Entomol. Bull. 33: 32-37.

Chittenden, F. H. 1922. The cocklebur billbug. Canad. Entomol. 54: 217-220.

Cobia, D., and D. E. Zimmer. 1978. Sunflowers -- production, pests, and marketing. Ext. Bull. 25: 73 pages. North Dakota State Univ., Fargo.

Cockerell, T. D. A. 1914. The entomology of Helianthus. Canad. Entomol. 47: 191-196.

Comstock, J. A. 1946. A few pests of sunflower in California. Bull. Southern California Acad. Sci. 45: 141-144.

Criddle, N. 1922. Beetles injurious to sunflowers in Manitoba. Canad. Entomol. 54: 97-99.

DePew, L. J. 1983. Sunflower moth (Lepidoptera: Pyralidae): oviposition and chemical control of larvae on sunflowers. J. Econ. Entomol. 76: 1164-1166.

Dorrell, D. G. 1978. Processing and utilization of oilseed sunflower, pages 407-440. In. J. F. Carter (ed.) Sunflower science and technology. Agron. Monogr. 19, 505 pages. Amer. Agron. Soc., Madison, WI.

Drake, C. J., and H. M. Harris. 1926. The flower webworm, a new flower pest. Trans. Iowa State Hort. 223-226.

Ehart, O. R. 1974. Biology and economic importance of Suleima helianthana (Riley) on sunflower cultivars, Helianthus annuus L., in the Red River Valley. Thesis. 92 pages. North Dakota State Univ., Fargo.

Fick, G. N. 1978. Breeding and genetics, pages 279-338. In. J. F. Carter (ed.) Sunflower science and technology. Agron. Monogr. No. 19, 505 pages. Amer. Soc. Agron., Madison, WI.

Fick, G. N., and G. E. Auwater. 1981. Resistance to the sunflower midge. Proc. Sunflower Forum and Res. Workshop. Page 18. Sunflower Assoc. Amer., Fargo, ND.

Fisher, W. S. 1924. A new species of Ataxia from the United States (Coleoptera: Cerambycidae). Canad. Entomol. 56: 253-254.

Foote, R. H. 1960. The species of the genus Neotephritis Hendel in America north of Mexico (Diptera: Tephritidae). J. New York Entomol. Soc. 68: 145-151.

Gaudet, M. D., and J. T. Schulz. 1981. Transmission of Phoma oleracea var. helianthi-tuberosi by the adult stage of Apion occidentale. J. Econ. Entomol. 74: 486-489.

Gaudet, M. D., and J. T. Schulz. 1984. Association of a sunflower fungal pathogen, Phoma macdonaldii, and a stem weevil, Apion occidentale (Coleoptera: Curculionidae). Canad. Entomol. 116: 1267-1273.

Gednalske, J. V., and D. D. Walgenbach. 1984. Influence of insecticide application timing on damage by Smicronyx fulvus and S. sordidus (Coleoptera: Curculionidae). J. Econ. Entomol. 77: 1545-1548.

Genung, W. G., and V. E. Green, Jr. 1979. Insect pests of
    sunflower in Florida. The Sunflower 5: 10-11, 29-31.
    Sunflower Assoc. Amer., Fargo, ND.
Genung, W. G., and V. E. Green, Jr. 1983. Two beetles new
    to the Everglades that attack sunflower (Coleoptera:
    Curculionidae and Cerambycidae). Fla. Entomol. 66:
    207-208.
Gershenzon, J., M. C. Rossiter, T. J. Mabry, C. E. Rogers,
    M. H. Blast, and T. L. Hopkins. 1985. Insect anti-
    feedant terpenoids in wild sunflower — a possible
    source of resistance to the sunflower moth. Pages
    433-446. In. P. A. Hedin (ed.) Bioregulatorors for
    pest control. ACS Symp. Ser. No. 276. Amer. Chem.
    Soc.
Hamilton, R. W. 1973. Observations on the biology of Hap-
    lorhynchites aeneus (Boheman) (Coleoptera: Curculioni-
    dae). Coleopt. Bull. 27: 83-86.
Hamilton, R. W. 1974. The genus Haplorhynchites (Coleop-
    tera: Rhynchitidae) in America north of Mexico. Ann.
    Entomol. Soc. Amer. 67: 787-794.
Hatchett, J. H., D. M. Daugherty, J. C. Robbins, R. M.
    Barry, and E. C. Houser. 1975. Biology in Missouri
    of Dectes texanus, a new pest of soybean. Ann. Ento-
    mol. Soc. Amer. 68: 209-213.
Hayes, W. P. 1917. Studies on the life-history of Ligyrus
    gibbosus DeG. (Coleoptera). J. Econ. Entomol. 10:
    253-260.
Heilman, T. J., J. V. Gednalske, and D. D. Walgenbach.
    1983. A simple washing method for extracting insect
    larvae from the soil. J. Kans. Entomol. Soc. 56:
    496-498.
Heinrich, C. 1921. Some Lepidoptera likely to be confused
    with the pink bollworm. J. Agric. Res. 20: 807-836.
Heiser, C. B., Jr. 1976. The sunflower. Univ. Okla.
    Press, Norman. 198 pages.
Heiser, C. B., Jr. 1985. The origin of domesticated sun-
    flowers. Pages 15-16. In International Sunflower
    Yearbook. 56 pages. Toowoomba, Australia.
Heiser, C. B., Jr. 1978. Taxonomy of Helianthus and ori-
    gin of domesticated sunflower, pages 31-53. In. J. F.
    Carter (ed.) Sunflower science and technology. Agron.
    19. Amer. Soc. Agron. 505 pages, Madison, WI.
Heiser, C. B., Jr., D. M. Smith, S. B. Clevenger, and W. C.
    Martin, Jr. 1969. The North American Sunflowers
    (Helianthus). Mem. Torrey Bot. Club 22: 218 pages.
Hilgendorf, J. H., and R. D. Goeden. 1981. Phytophagous
    insects reported from cultivated and weedy varieties

of the sunflower, <u>Helianthus</u> <u>annuus</u> L., in North America. ESA Bull. 27: 102-108.

Johnson, A. L., and B. H. Beard. 1977. Sunflower moth damage and inheritance of the phytomelanin layer in sunflower achenes. Crop Sci. 17: 369-372.

Kamali, K. 1973. Biology and ecology of the tephritid complex on sunflower (Diptera: Tephritidae). Thesis. 128 pages. North Dakota State Univ., Fargo.

Kamali, K., and J. T. Schulz. 1974. Biology and ecology of <u>Gymnocarena</u> <u>diffusa</u> (Diptera: Tephritidae) on sunflower in North Dakota. Ann. Entomol. Soc. Amer. 67: 695-699.

Khaemba, B. M., and M. J. Mutinga. 1982. Insect pests of sunflower (<u>Helianthus</u> <u>annuus</u> L.) in Kenya. Insect Sci. Applic. 3: 281-286.

Klisiewicz, J. M. 1979. Relation of infestation with sunflower moth <u>Homoeosoma</u> <u>electellum</u> larvae to the incidence of <u>Rhizopus</u> rot in sunflower seed heads. Canad. J. Plant Sci. 59: 797-801.

Kopp, D. D. 1983. 1982 midge damage survey. Proc. Sunflower Workshop. Page 20. National Sunflower Assoc. Fargo, ND.

Kreitner, G., and C. E. Rogers. 1981. Sunflower seed midge: effects of larval infestation on pericarp development in sunflower. Ann. Entomol. Soc. Amer. 74: 431-435.

Laferriere, J. E. 1986. Interspecific hybridization in sunflowers: an illustration of the importance of wild genetic resources in plant breeding. Outlook on Agric. 15: 104-109.

Lofgren, J. R. 1983. Damage to hybrids and inbreds inflicted by sunflower midge (<u>Contarinia</u> <u>schulzi</u> Gagne). Proc. Sunflower Res. Workshop. Page 21. National Sunflower Assoc., Fargo, ND.

Lynch, R. E., and J. W. Garner. 1980. Insects associated with sunflower in south Georgia. J. Ga. Entomol. Soc. 15: 182-189.

McBride, D. K., L. D. Charlet, and C. Y. Oseto. 1985. Banded sunflower moth. Ext. Leaflet E-823, 3 pages. North Dakota State Univ., Fargo.

McBride, D. K., D. D. Kopp, C. Y. Oseto, and J. D. Busacca. 1981. Insect pest management for sunflower. Ext. Bull. 28: 23 pages. North Dakota State Univ., Fargo.

McGregor, S. E. 1976. Insect pollination of cultivated crop plants. Agric. Handbook 496: 411 pages. USDA, ARS, Washington, D.C.

Muma, M. H., R. N. Lyness, C. E. Classen, and A. Hoffman.
    1950. Control tests on sunflower insects in Nebraska.
    J. Econ. Entomol. 43: 477-480.

Neck, R. W. 1973. Food plant ecology of the butterfly
    Chlosyne lacinia (Geyer) (Nymphalidae) I. Larval food
    plants. J. Lepidopt. Soc. 27: 22-23.

Neck, R. W. 1977. Food plant ecology of the butterfly
    Chlosyne lacinia (Geyer) (Nymphalidae). II. Additional
    larval food plant data. J. Res. Lepidopt. 16: 69-74.

Newton, J. H. 1921. A sunflower pest. Colorado State
    Entomol. Soc. Circ. 34: 37-38.

Orfila, R. N. 1964. Una plaga nueva para la Argentina: la
    "isoca espinosa del girasol." Inf. de Invest. Agric.
    193: 41-48.

Oseto, C. Y., and G. A. Braness. 1979. Bionomics of Smi-
    cronyx fulvus (Coleoptera: Curculionidae) on culti-
    vated sunflower, Helianthus annuus. Ann. Entomol.
    Soc. Amer. 72: 524-528.

Oseto, C. Y., and G. A. Braness. 1980. Chemical control
    and bioeconomics of Smicronyx fulvus on cultivated
    sunflower in North Dakota. J. Econ. Entomol. 73:
    218-220.

Oseto, C. Y., and L. D. Charlet. 1981. Soil distribution
    of overwintering Smicronyx fulvus (Coleoptera: Curcul-
    ionidae) larvae in North Dakota. J. Kans. Entomol.
    Soc. 54: 784-786.

Oseto, C. Y., W. F. Burr, and R. B. Carlson. 1982. Rela-
    tionship of sunflower planting dates to damage by
    Cylindrocopturus adspersus (Coleoptera: Curculioni-
    dae) in North Dakota. J. Econ. Entomol. 75: 761-
    764.

Oursbourn, C. D., W. A. Lepori, R. D. Lacewell, K. Y. Lam,
    and O. B. Schacht. 1978. Energy potential of Texas
    crops and agricultural residues. Texas Agric. Exp.
    Sta. Misc. Publ. 1361. 36 pages. Texas A&M Univ.,
    College Station.

Parker, F. D. 1981. Sunflower pollination: Abundance,
    diversity, and seasonality of bees and their effect on
    seed yield. J. Agric. Res. 20: 49-61.

Phillips, R. L., N. M. Randolph, and G. L. Teetes. 1973.
    Seasonal abundance and nature of damage of insects
    attacking cultivated sunflowers. Texas Agric. Exp.
    Sta. Misc. Publ. MP-1116, 7 pages. Texas A&M Univ.,
    College Station.

Poston, F. L., R. B. Hammond, and L. P. Pedigo. 1977.
    Growth and development of the painted lady on soybeans

(Lepidoptera: Nymphalidae). J. Kans. Entomol. Soc. 50: 31-36.

Putt, E. D. 1978. History and present world status. Pages 1-29. In J. F. Carter (ed.) Sunflower science and technology. Agron. Monogr. 19, 505 pages. Amer. Agron. Soc., Madison, WI.

Rajamohan, N. 1976. Pest complex on sunflower -- a bibliography. PANS 22: 546-563.

Robertson, C. 1922. The sunflower and its insect visitors. Ecology 3: 17-21.

Rogers, C. E. 1974. Bionomics of the carrot beetle in the Texas Rolling Plains. Environ. Entomol. 3: 969-974.

Rogers, C. E. 1977a. Cerambycid pests of sunflower: distribution and behavior in the southern Plains. Environ. Entomol. 6: 833-838.

Rogers, C. E. 1977b. Bionomics of the sunflower beetle. Environ. Entomol. 6: 466-468.

Rogers, C. E. 1977c. Hosts and parasitoids of the Cecidomyiidae (Diptera) in the Rolling Plains of Texas. J. Kans. Entomol. Soc. 50: 179-186.

Rogers, C. E. 1978. Sunflower moth: feeding behavior of the larva. Environ. Entomol. 7: 763-765.

Rogers, C. E. 1979a. Selected bibliography of insect pests of sunflower. Texas Agric. Exp. Sta. Misc. Publ. MP-1439, 41 pages. Texas A&M Univ., College Station.

Rogers, C. E. 1979b. Sunflower budmoth: behavior and impact of the larva on sunflower seed production in the southern Plains. Environ. Entomol. 8: 113-116.

Rogers, C. E. 1980a. Biology and breeding for insect and disease resistance in oilseed crops, pages 359-389. In. M. K. Harris (ed.) Biology and breeding for resistance to arthropods and pathogens in agricultural plants. Texas Agric. Exp. Sta. Misc. Publ. MP-1451, 605 pages. Texas A&M Univ., College Station.

Rogers, C. E. 1980b. Natural enemies of insect pests of sunflower: a world view. Texas Agric. Exp. Sta. Misc. Publ MP-1457. 30 pages. Texas Agric. Exp. Sta., College Station.

Rogers, C. E. 1981a. Breeding sunflower for resistance to insects and diseases in the United States. Proc. EUCARPIA, pages 175-213. Prague, Czechoslovakia.

Rogers, C. E. 1981b. Resistance of sunflower species to the western potato leafhopper. Environ. Entomol. 10: 697-700.

Rogers, C. E. 1982. The midge bestiary. The Sunflower 8:35. Sunflower Assoc. Amer., Fargo, ND.

Rogers, C. E. 1985a. Extrafloral nectar: entomological implications. ESA Bull. 15-20.

Rogers, C. E. 1985b. Bionomics of _Eucosma womonana_ Kearfott (Lepidoptera: Tortricidae), a root borer of sunflower. Environ. Entomol. 14: 42-44.

Rogers, C. E. 1985c. Cultural management of _Dectes texanus_ (Coleoptera: Cerambycidae) in sunflower. J. Econ. Entomol. 78: 1145-1148.

Rogers, C. E., T. L. Archer, and E. D. Bynum, Jr. 1984a. _Bacillus thuringiensis_ for controlling larvae of _Homoeosoma electellum_ on sunflower. J. Agric. Entomol. 1: 323-329.

Rogers, C. E., A. P. Arthur, and D. J. Bauer. 1986a. Long range migration by the sunflower moth. Pages 3-9. _In._ A. N. Sparks (ed.) Long-range migration of moths of agronomic importance to the United States and Canada. USDA, ARS Series 43. 104 pages. U.S. Governemnt Print. Office, Wash., DC.

Rogers, C. E., J. Gershenzon, N. Ohno, T. J. Mabry, R. D. Stipanovic, and G. L. Kreitner. 1987. Terpenes of wild sunflower (_Helianthus_): an effective mechanism against seed predation by larvae of the sunflower moth, _Homoeosoma electellum_ (Hulst) (Lepidoptera: Pyralidae). Environ. Entomol. (in press).

Rogers, C. E., and G. R. Howell. 1973. Behavior of the carrot beetle on native and treated commercial sunflower. Texas. Agric. Exp. Sta. Prog. Rept. PR-3249, 7 pages. Texas A&M Univ., College Station.

Rogers, C. E., H. B. Jackson, R. D. Eikenbary, and K. J. Starks. 1972. Host-parasitoid interaction of _Aphis helianthi_ on sunflowers with introduced _Aphelinus asychis_, _Ephedrus plagiator_, and _Praon volucre_, and native _Aphelinus nigritus_ and _Lysiphlebus testaceipes_. Ann. Entomol. Soc. Amer. 65: 38-41.

Rogers, C. E., and O. R. Jones. 1979. Effects of planting date and soil water on infestation by larvae of _Cylindrocopturus adspersus_. J. Econ. Entomol. 72: 529-531.

Rogers, C. E., and G. L. Kreitner. 1983. Phytomelanin of sunflower achenes: a mechanism for pericarp resistance to abrasion by larvae of the sunflower moth (Lepidoptera: Pyralidae). Environ. Entomol. 12-277-285.

Rogers, C. E., W. D. Perkins, and G. J. Seiler. 1986b. Bionomics of _Stibadium spumosum_ Grote (Lepidoptera: Noctuidae), a pest of sunflower (_Helianthus annuus_ L.) in the southern Plains. Environ. Entomol. 15: (in press).

Rogers, C. E., and G. J. Seiler. 1985. Sunflower (Helianthus) resistance to a stem weevil, Cylindrocopturus adspersus (Coleoptera: Curculionidae). Environ. Entomol. 14: 624-628.

Rogers, C. E., and J. G. Serda. 1982. Cylindrocopturus adspersus in sunflower: overwintering and emergence patterns on the Texas High Plains. Environ. Entomol. 11: 154-156.

Rogers, C. E., and T. E. Thompson. 1978a. Evaluation of Helianthus for resistance to insect pests. Proc. 8th Internat'l. Sunflower Assoc.: 320-327. Internat'l. Sunflower Assoc.

Rogers, C. E., and T. E. Thompson. 1978b. Helianthus resistance to the carrot beetle. J. Econ. Entomol. 71: 760-761.

Rogers, C. E., and T. E. Thompson. 1978c. Resistance in wild Helianthus to the sunflower beetle. J. Econ. Entomol. 71: 622-623.

Rogers, C. E., and T. E. Thompson. 1978d. Resistance of wild Helianthus species to an aphid, Masonaphis masoni. J. Econ. Entomol. 71: 221-222.

Rogers, C. E., and T. E. Thompson. 1979. Helianthus resistance to Masonaphis masoni. The Southwest. Entomol. 4: 321-324.

Rogers, C. E., and T. E. Thompson. 1980. Helianthus resistance to the sunflower beetle. J. Kans. Entomol. Soc. 53: 727-730.

Rogers, C. E., T. E. Thompson, and R. J. Gagne. 1979a. Cecidomyiidae of Helianthus: taxonomy, hosts, and distribution. Ann. Entomol. Soc. Amer. 72: 109-113.

Rogers, C. E., T. E. Thompson, and O. R. Jones. 1979b. Eucosma womonana Kearfott (Lepidoptera: Oleuthreutidae): a new pest of sunflower in the southern Plains. J. Kans. Entomol. Soc. 52: 373-376.

Rogers, C. E., T. E. Thompson, and G. J. Seiler. 1982. Sunflower species of the United States. 75 pages. National Sunflower Assoc., Bismarck, ND.

Rogers, C. E., T. E. Thompson, and M. B. Stoetzel. 1978a. Aphids of sunflower: distribution and hosts in North America (Homoptera: Aphididae). Proc. Entomol. Soc. Wash. 80: 508-513.

Rogers, C. E., T. E. Thompson, and G. J. Seiler. 1984b. Registration of three Helianthus germplasm for resistance to the sunflower moth. Crop. Sci. 24: 212-213.

Rogers, C. E., T. E. Thompson, and M. J. Wellik. 1980. Survival of Bothynus gibbosus (Coleoptera: Scarabaeidae) on _Helianthus_ species. J. Kans. Entomol. Soc. 53: 490-494.

Rogers, C. E., T. E. Thompson, and D. E. Zimmer. 1978d. _Rhizopus_ head rot of sunflower: etiology and severity in the southern Plains. Plant Dis. Rept. 62: 769-771.

Rogers, C. E., and E. W. Underhill. 1983. Seasonal flight for the sunflower moth (Lepidoptera: Pyralidae) on the Texas High Plains. Environ. Entomol. 12: 252-254.

Rogers, C. E., P. W. Unger, T. L. Archer, and E. D. Bynum, Jr. 1983. Management of a stem weevil, _Cylindrocopturus adspersus_ in sunflower in the southern Great Plains. J. Econ. Entomol. 76: 952-956.

Rossiter, M. C., J. Gershenzon, and T. J. Mabry. 1986. Behavioral and growth responses of specialist herbivore, _Homoeosoma electellum_ to major terpenoid of its host, _Helianthus_ spp. J. Chem. Ecol. 12: 1505-1521.

Satterthwait, A. F. 1946. Sunflower seed weevils and their control. J. Econ. Entomol. 39: 787-792.

Satterthwait, A. F. 1948. Important sunflower insects and their insect enemies. J. Econ. Entomol. 41: 725-731.

Schrader, W. 1921. Experiments on a species of migrating butterfly. Bull. So. Calif. Acad. Sci. 27: 68-70.

Schulz, J. T. 1973. Damage to cultivated sunflower by _Contarinia schulzi_. J. Econ. Entomol. 66: 282.

Schulz, J. T. 1978. Insect pests, pages 169-223. _In_. J. F. Carter (ed.) Sunflower science and technology. Agron. Monogr. 19. 505 pages. Amer. Soc. Agron., Madison, WI.

Seiler, G. J. 1985. Evaluation of seeds of sunflower species for several chemical and morphological characters. Crop Sci. 25: 183-187.

Seiler, G. J., R. E. Stafford, C. E. Rogers. 1984. Prevalence of phytomelanin in pericarps of sunflower parental lines and wild species. Crop Sci. 24: 1202-1204.

Shapiro, I. D. 1975. Achievements of plant breeding in plant resistance to pests. Proc. 8th Internat'l. Plant Prod. Congr. 3: 162-167. Moscow, USSR.

Stafford, R. E., C. E. Rogers, and G. J. Seiler. 1984. Pericarp resistance to mechanical puncture in sunflower achenes. Crop Sci. 24: 891-894.

Stamp, N. 1977. Aggregation behavior of _Chlosyne lacinia_ larvae (Nymphalidae). J. Lepidopt. Soc. 31: 35-40.

Stipanovic, R. D., D. H. O'Brien, C. E. Rogers, and K. D. Hanlon. 1980. Natural rubber from sunflower. J. Agric. Food Chem. 28: 1322-1323.

Stipanovic, R. D., G. J. Seiler, and C. E. Rogers. 1982. Natural rubber from sunflower. 2. J. Agric. Food Chem. 30: 611-613.

Stoenescu, F., A. Ulinici, H. Illiescu, and F. Paulian. 1974. Pests. Pages 296-307. In. Sunflower. 332 pages. Acad. Rom. Soc. Rep.

Teetes, G. L., and N. M. Randolph. 1968. Chemical control of the sunflower moth on sunflowers. J. Econ. Entomol. 61: 1344-1347.

Teetes, G. L., and N. M. Randolph. 1969a. Seasonal abundance and parasitism of the sunflower moth, Homoeosoma electellum, in Texas. Ann. Entomol. Soc. Amer. 62: 1461-1464.

Teetes, G. L., and N. M. Randolph. 1969b. Chemical and cultural control of the sunflower moth in Texas. J. Econ. Entomol. 62: 1444-1447.

Teetes, G. L., and N. M. Randolph. 1969c. Some new host plants of the sunflower moth in Texas. J. Econ. Entomol. 62: 264-265.

Teetes, G. L., and N. M. Randolph. 1971. Differences in susceptibility of certain sunflower varieties and hybrids to the sunflower moth. J. Econ. Entomol. 64: 1285-1287.

Thompson, T. E., C. E. Rogers, and D. Z. Zimmerman. 1980. Sunflower oil quality and quantity as affected by Rhizopus head rod. J. Amer. Oil Chem. Soc. 57: 106-108.

Thompson, T. E., D. C. Zimmerman, and C. E. Rogers. 1981. Wild Helianthus as a genetic resource. Field Crops Res. 4: 333-343.

Underhill, E. W., A. P. Arthur, M. D. Chisholm, and W. F. Steck. 1979. Sex pheromone componenets of the sunflower moth, Homoeosoma electellum: Z-9, E-12-tetradecadienol and Z-9-tetradedicenol. Environ. Entomol. 8: 740-743.

Underhill, E. W., A. P. Arthur, and P. G. Mason. 1986. Sex pheromone of the banded sunflower moth, Cochylis hospes (Lepidoptera: Cochylidae): identification and field trapping. Environ. Entomol. 15: 1063-1066.

Underhill, E. W., C. E. Rogers, and L. R. Hogge. 1987. Sex attractants for two sunflower pests, Eucosma womonana (Lepidoptera: Tortricidae) and Isophrictis similiella (Lepidoptera: Gelechiidae). Environ. Entomol. 16: (in press).

Vaurie, P. 1981. Revision of Rhodobaenus. Part 2. Species in North America (Canada to Panama) (Coleoptera,

Curculionidae, Rhynchophorinae). Bull. Amer. Mus. Natr'l. Hist. 171: 122-174.

Vinall, H. N. 1922. The sunflower silage crop. USDA Bull. 1045.

Waiss, A. C., Jr., B. G. Chan, C. A. Elliger, V. H. Garrett, E. C. Carlson, and B. H. Beard. 1977. Larvicidal factors contributing to the host plant resistance against sunflower moth. Naturwissens 64: 341.

Walker, F. H., Jr. 1936. Observations on sunflower insects in Kansas. J. Kans. Entomol. Soc. 9: 16-25.

Weiss, H. B., and R. B. Lott. 1923. Notes on *Rhodobaenus 13-punctatus* (Ill.), the cocklebur bill-bug. (Col.) Entomol. News 34: 103-106.

Westdal, P. H. 1949. Life history, biology, and control of *Phalonia hospes* Wlshm. in Manitoba. Tech. Rep. Agric. Can. Res. Sta. 31 pages. Winnipeg, Manitoba.

Westdal, P. H., and C. F. Barrett. 1960. Life-history and hosts of the sunflower maggot, *Strauzia longipennis* (Weid.) (Diptera: Tryptidae), in Manitoba. Canad. Entomol. 92: 481-488.

Westdal, P. H., and C. F. Barrett. 1962. Injury by the sunflower maggot, *Strauzia longipennis* (Wied.) (Diptera: Tryptidae), to sunflowers in Manitoba. Canad. J. Plant Sci. 42: 11-14.

Williams, C. B. 1970. The migrations of the painted lady butterfly, *Vanessa cardui* (Nymphalidae), with special reference to North America. J. Lepidopt. Soc. 24: 157-175.

Worth, C. B. 1943. Ecologically interrelated insects on a sunflower (Hymenoptera, Formicidae, Ichneumonidae, and Chalcidoidae; Hemiptera, Membricidae, and Aphididae; Diptera, Syrphidae). Entomol. News 54: 156-163.

Yang, S. M., and D. F. Owen. 1982. Symptomatology and detection of *Macrophomina phaseolina* in sunflower plants parasitized by *Cylindrocopturus adspersus*. Phytopathology 72: 819-821.

Yang, S. M., C. E. Rogers, and D. N. Luciani. 1983. Transmission of *Macrophomina phaseolina* in sunflower by *Cylindrocopturus adspersus*. Phytopathology 73: 1467-1469.

Zimmer, D. E., and J. A. Hoes. 1978. Diseases. Pages 225-262. In. J. F. Carter (ed.) Sunflower Science and Technology. Agron. Monogr. 19, 505 pages. Amer. Agron. Soc., Madison, WI.

2

Entomology of Crucifers and Agriculture— Diversification of the Agroecosystem in Relation to Pest Damage in Cruciferous Crops

S. Finch

Considerable work has already been published on the community organization of the complete complex of insects found on cultivated and wild cruciferous plants (Pimentel 1961a & b, Root and Tahvanainen 1969, Tahvanainen 1972, Root 1973), and therefore this broad approach will not be adopted in this review. Instead I will follow the example of van Emden (1965) and concentrate on just one species, in this case the root-infesting cabbage root fly, *Delia radicum* (L.) (Diptera: Anthomyiidae), to develop a theme based largely on data from behavioural studies. By concentrating in this manner, I hope to produce a critical rather than an encyclopaedic review. My main aim will be to describe how cruciferous weeds and diversified crop habitats influence the pest status of this insect.

Unlike other reviews on this subject, I will devote considerable time to the primary producer, the plant, since this has a major influence on the subsequent insect/host-plant interactions. I will concentrate mainly on how weeds influence the overall levels of pest infestation, and whether, as suggested many times in the past (eg. Ullyett 1947, Elton 1958), hedgerows are reservoirs for predators and parasites of insect pests of crops. I will conclude by reviewing some of the systems now being proposed for alleviating pest problems in cruciferous crops.

CHEMICAL FACTORS CHARACTERIZING CRUCIFEROUS PLANTS

During evolution, plants have been subjected to a variety of pressures not only from the climate and soil type, but also from competing plant species and herbivores.

The evolution of interactions between plants and insects frequently involves a "leap-frogging" process in which at any one time selection may favour either the plant or the insect (Finch 1980). A plant's best strategy against being eaten is to become inedible (Dethier 1970). Chemicals that confer inedibility on a plant belong to a wide range of organic compounds, generally classed as "secondary plant substances" (Czapek 1922). Plants of the family Cruciferae are characterized by secondary plant substances known as glucosinolates (mustard oil glycosides), which when hydrolyzed, either naturally or when plant tissues are damaged, yield glucose, sulphate and the corresponding mustard oils many of which are volatile isothiocyanates (Kjaer 1963). Plants protected from the attacks of phytophagous insects by such chemicals have in a sense entered a new niche (Ehrlich and Raven 1964). If, however, a mutation or recombinant appears in an insect population that enables individuals to feed on such plants, selection favours survival of those individuals.

As a result of coevolution, therefore, each plant species has come to harbour a more or less distinct arthropod fauna (Ehrlich and Raven 1964, Whittaker and Feeny 1971) which, because they now restrict themselves to limited food plants, are referred to as "specialists". Should host plants be present in only small numbers, 'specialist' feeders must allocate considerable time and energy in searching for them. Consequently, ephemeral plants such as the Cruciferae rely for survival on escape from their specialist feeders (Root 1973) and produce energetically inexpensive toxins (e.g. gluconsinolates) to minimize defoliation by generalist feeders (Feeny 1976).

ECOLOGICAL FACTORS CHARACTERIZING CRUCIFEROUS PLANTS

The Cruciferae are described as a family of mostly annual (some biennial) or ephemeral herbs of open habitats and, because of this trait, are common weeds of cultivation (Clapham, Tutin and Warburg 1958). There are a few such as Armoracia and Hesperis that are perennial and can compete in undisturbed ground, but the vast majority cannot.

Most genera are adapted to the niche provided by unstable land surfaces and many (e.g. spp of Crambe, Cakile, Matthiola, Brassica, Raphanus) are associated with the coast, where fresh soil surfaces, often mainly sub-soil, are produced by erosion. The relatively mild climates along the coasts of western Europe and the

Mediterranean appear to be the main reason for their success, since the biennial species remain green and carry on growing through the winter. In this way they cover ground with their basal rosettes, and make space for themselves prior to the start of the new growing season of their competitors. Apart from the seashore and cultivated land, crucifers are found frequently alongside streams and rivers where there are disturbed areas of soil similar to those along the coast. In addition to genera that grow alongside water, some crucifers live in dry walls (e.g. _Arabis_ spp), others are true alpines (e.g. _Draba_ spp), an occasional one (e.g. _Dentaria bulbifera_) is found in woodlands and others (e.g. _Barbarea_ & _Alliaria_) are associated with hedgerows. The latter, however, generally only colonize areas where the soil has been disturbed by animal activity.

Another way Cruciferae differ from many other plant families is they lack mycorrhiza (Hardman - personal communication). Mycorrhiza are generally present in plant families that grow in undisturbed high-humus soils like those found in woodlands and hedgerows. Mycorrhiza presumably are of little benefit to cruciferous plants whose strong roots have adapted to exploit the small amounts of nutrients in raw soils lacking humus. Hence coastal types of cruciferous plants are well-adapted to grow on cultivated soils lacking structure and humus. Perhaps as a consequence, they have responded well to cultivation and have been used to produce many of our existing vegetable crops.

Apart from biennial crucifers being able to create their own space, annual cruciferous weeds have space provided for them by land cultivation and by the removal of other weed competitors by selective herbicides. Obviously, herbicides cannot be effective against cruciferous plants in brassica crops or they would kill the crop. As a consequence, many cruciferous weeds set seed in brassica crops and such seed can remain viable for many years, germinating only when it is brought near to the surface by burrowing animals or by man. The large numbers of plants of _Sinapis arvensis_ that appear on soil banks during road construction are a good example of this (Roberts 1982).

FACTORS CONTRIBUTING TO INSECT PREVALENCE IN CROPS

Locating new host-plants appears to be one of the major factors regulating pest insect populations. The prevalence of insects in crops is partly due to coevolution but mainly due to the activities of man. In general, modern crop production creates large ecological disturbances, since many cruciferous crops are in the ground for only a small part of each year. As a consequence, new habitats are continually arising. If, therefore, an insect species is to exploit such a habitat successfully, it must have a level of dispersal geared to the rate of change of its habitat (Southwood 1962). As a result of selection, insects of temporary habitats are now usually highly mobile and most appear to have an "innate tendency to disperse" (Andrewartha and Birch 1954). This does not necessarily mean, however, that all of the insects in a population will disperse. There are obvious biological advantages for any population if individuals have differing potentials for dispersal. For example, Wellington (1957) suggested that when the members of certain populations emerge, most of the strongest fliers disperse before ovipositing, but some oviposit locally. In contrast, the less active adults are incapable of sustained flight, so that all weak females, many of which may have mated with the active males, oviposit near their site of emergence. Thus partial emigration of the better stock coupled with complete retention of the poorer, but still viable, stock hastens the decline of population quality in a locality, yet at the same time makes maximum use of the available resources.

Many of the crop plants cultivated by man are essentially 'unapparent', since they are derived from early successional herbs (Feeny 1976), but through monoculture their apparency to insects has been greatly increased. Consequently, following the continuous selection processes, many crop plants are now highly preferred by insects. It is not surprising, therefore, that since many crop plants are now apparent but lack the appropriate biochemical defences commensurate with such a state (Feeny 1976), insecticides must be applied to prevent widespread damage by adapted phytophagous insects.

Like its major cultivated host-plant _Brassica_, the cabbage root fly also appears to have made the transition from the coast to inland Britain. Of eighty-six cruciferous plant species inoculated with root fly eggs in the laboratory, all of the coastal species supported the pest, and genera like _Cochlearia_ were so nutritive that the insect could complete its development on exceptionally small amounts of root tissue (Finch and Ackley 1977). In addition to species of <u>Cochlearia</u>, coastal species of <u>Brassica</u>, <u>Raphanus</u> and <u>Barbarea</u> were also good host plants.

INSECT HOST-PLANT RELATIONSHIPS

The selection of a host-plant by an insect is not just a series of 'take it or leave it' situations but is a finely tuned system based on the physiological state of both the insect and the plant. Therefore, if comparisons are to be made between plants growing under different conditions (e.g. Pimentel 1961), the two sets of plants must have similar rates of growth. The relationship between an insect and its host-plant is often extremely complicated, since factors regulating host plant selection by the adult, and subsequent survival of its offspring, vary considerably throughout the season.

For example, natural populations of cruciferous seedlings often emerge in dense stands which have survival value merely as a result of their numbers (Pimentel 1961). Another factor in this survival, however, is that the secondary plant chemicals often fall, during weeks 1-4 of seedling growth, to extremely low levels (Cole 1980). This implies that although such seedlings may be visually apparent they may be much less chemically apparent to the specialist herbivores trying to locate them.

Plant type is also extremely important during host-plant selection (Hardman and Ellis, 1978). However, irrespective of where cabbage root fly populations were collected in the British Isles during 1976-1979, all showed a stronger preference for cultivars of <u>Brassica napus</u> (swede) than for cultivars of <u>B. oleracea</u> (cabbage, cauliflower or Brussels sprouts) (Finch 1980).

The isothiocyanates and related compounds are the characteristic chemicals that the specialist insects use to locate cruciferous host-plants (Verschaffelt 1910). The concentration of free isothiocyanates in any intact tissue is negligible; isothiocyanates are formed only after injury. It can be argued, however, that such chemicals are being produced continuously during the rigorous cell transformations involved in plant growth (Schraudolf and Bergmann 1965). Such transformations are likely to be at maximum in those plant species or cultivars which grow rapidly and those containing attractant chemicals should be among the most attractive to pests.

This is supported by laboratory experiments. The fast-growing plants <u>Brassica rapa</u> var. <u>rapa</u> (L.) Thell,

<u>Sinapis</u> <u>alba</u> L. and <u>Brassica</u> <u>chinensis</u> L. were the most preferred by the cabbage root fly (Finch et al. 1975), the diamond-back moth, <u>Plutella</u> <u>xylostella</u> (Price 1975) and the mustard beetle, <u>Phaedon</u> <u>cochleariae</u> (F.), (Finch and Gairn 1976). It must always be remembered, however, that a chemical analysis only detects the quantities of chemicals present at the time of analysis and does not reveal how rapidly they are being used or replaced within the intact plant tissues.

The importance of growth rate was established by comparing root fly oviposition on different age plants of the same cauliflower cultivar all growing in similar-sized pots. Initially, the largest plants were preferred but once such plants became pot-bound they stayed the same size and the next younger plants of a similar size but still actively-growing became preferred, with the preference gradually moving back to the next younger plants the longer the experiment was continued. Transplanting into larger pots at the appropriate time, prevented this check in growth and then the oldest, and largest, plants remained the most preferred throughout. As with most areas of host-plant selection, the visual, olfactory and gustatory stimuli all interact in the insects' final choice. Chemical stimulation, however, plays a major role in host-plant acceptance since, when forty cauliflower plants of one cultivar were selected from a batch of 400 for their apparent visual similarity, approximately sixteen times as many eggs were laid on the most preferred as on the least preferred plant.

One point to be wary of when studying insect/host-plant relationships is the widely-held belief that when you concentrate on just one insect species, the system is simple. Admittedly the system is less complicated, but this is not necessarily the same as being simple. It is now well-documented that chemicals released by the host-plant have a dramatic influence on host-plant selection by the cabbage root fly (Finch, 1980). There are also qualitative differences in the glucosinolates present in different plant species (Kjaer 1963, Hicks 1974, Finch et al. 1975) and even within one plant, these chemicals vary with age (Chong and Bible 1974, Ellis et al. 1980), plant part (Joseffson 1967) and sulphur nutrition (Freeman and Mossadeghi 1972).

Of the species of Cruciferae native to the British Isles, 83 were tested both under laboratory and field-cage conditions to determine which stimulated most oviposition by cabbage root fly. It was deduced that ten chemicals

(Finch 1977) could be involved in the process of host-plant selection by this fly. The quantities of some of these chemicals changed during the life of the plant, however, and certain 'attractive' chemicals eventually concentrated sufficiently to become repellent. For example, flies oviposited on wild radish, Raphanus raphanistrum L., in which the concentrations of allylisothiocyanate never exceeded 11 μg/g, but not on either B. nigra or Brassica juncea (L.) Czern and Cross, in which the minimum concentration was 25 μg/g. It was also clear that certain chemicals were moved from one site to another during plant growth, since attractants were present in the foliage of young plants but not of old plants of Sinapis alba L. and Sisymbrium orientale L. By the time the old plants were tested, however, they had started to seed. Subsequently, it was shown that volatile chemicals similar to those found previously in the leaves were present in the seeds. The removal of such chemicals from the photosynthetic tissue of the senescing plants to the seeds obviously favours survival of the plant species. This finding also suggests that ephemeral plants in which seed can be produced within weeks of the plant germinating are likely to have an extremely high flux of protective chemicals and hence may not be as vulnerable to being eaten by insects as is sometimes suggested. The movement of 'protective' chemicals to the seed is also of survival value to insects, since it prevents eggs being laid on senescing plants that would probably not live long enough to support insect larval development.

In addition to the plant affecting host-plant selection, the early colonizing insects can also alter the attractiveness of the plant to would-be later colonizers. In the field there are usually certain plants, the most actively-growing, on which the female flies prefer to lay. Once the eggs hatch and the larvae begin to feed on the roots, however, the chosen plants show signs of stress and soon lose their preferred status. Other plants then take over this role. As a consequence, by the end of a generation oviposition may appear to have been regular across a field whereas by recording at more frequent intervals it is only too obvious that it was not. Larvae feeding on the roots of brassica plants also cause such plants to develop a reddish hue. In this condition, the plants not only become less attractive chemically but also less attractive visually (Prokopy, Collier and Finch 1983).

The rate of plant growth is even more important to insects that produce subsequent generations on the host-plant selected by the original colonizers. Kennedy, Booth and Kershaw (1959) showed that the aphid _Brevicoryne brassicae_ preferred the newly-produced young leaves of brassica plants, and that the rate of plant growth affected not only the number of aphids alighting on such plants but also regulated the subsequent fecundity of those that settled. Hence a factor that influences plant growth can have a considerable effect on host-plant selection by insects. It cannot be summarily dismissed, or adjusted mathematically, by regarding smaller plants growing in weedy situations as being qualitatively similar to their larger counterparts growing in the absence of competition from other plant species.

There are many other factors that help to stabilize the overall population of a particular crop pest. Not least is the fact that plant species highly preferred by the adults for oviposition are not always the species most nutritive as food for the larvae, and vice versa. There is also a considerable partitioning of resources between the various pest species, such that within a crop type certain cultivars may be most preferred by certain pests, least preferred by others and of intermediate status to yet others (Radcliffe and Chapman 1966). Also the strategies used by the colonizing insects differ considerably. For example, _Pieris brassicae_ lays its eggs in large batches on just one or two plants, whereas _P. rapae_ lays a single egg on many of the plants it visits and seems to require an inter-oviposition bout of flying between laying one egg and the next. As both strategies are successful, the advantages of being conspicuous and feeding communally appear comparable to those of remaining relatively inconspicuous but having to feed alone.

PEST POPULATIONS ON NON-CULTIVATED CRUCIFEROUS PLANTS

The factor that seems to be overlooked most often when considering cruciferous wild-host plants as possible reservoirs of pest insects is the relative scarcity of such plants in localities devoted to vegetable production. Anyone who has searched for such plants will soon realize that cruciferous plants do not abound in hedgerows. This was clearly demonstrated in the study of van Emden (1965), who had to sow mustard (_Brassica nigra_) as an alternative host-plant alongside his experimental plots to arrest dispersing _Brevicoryne brassicae_ prior to the introduction of his experimental crop plants. It is imperative, therefore, in experiments to determine the relative distribution of pests on a range of host-plants, to ensure that all plants are available for selection when the pest population migrates to find new host-plants. With many species of aphid, once a particular plant has been chosen by the initial colonizer the plant is used as the source of food for many subsequent generations.

Findings of this type may help to explain why Root (1973) found no <u>Brevicoryne brassicae</u> infesting his 1967 collard crops. Perhaps the crop was transplanted too late? Admittedly it was planted on the same date (June 5) as the 1968 crop which did become infested, but calendar date is not the important factor in this context. Instead some form of physiological time scale has to be used to ensure that crops in subsequent years do coincide with the same periods of insect activity by the early-season colonizers.

The scarcity of crucifers in the vicinity of his experimental plots was also stressed by Root (1973). He showed that although the unmowed vegetation surrounding his experimental gardens consisted of grasses and herbs mixed in roughly equal proportions, yellow rocket (<u>Barbarea vulgaris</u>) was the only crucifer present, and even this species was distributed only in small scattered clumps. In contrast, other crucifers (<u>Brassica nigra</u>, <u>Capsella bursa-pastoris</u>, <u>Lepidium</u> spp.) were found at the edge of a cultivated field 100m from the gardens.

During the past 20 years at Wellesbourne, cruciferous weeds have been found almost exclusively in cultivated fields and predominantly in those growing brassica crops. As stated earlier, cruciferous weeds are restricted largely to disturbed soil situations and survive in greater numbers at the edge of, but within, the crop, largely because most cultural machinery is inoperative in the area of the crop used for turning.

In arable soils, an appreciable seed bank usually accumulates from weeds allowed to seed. A weedy field may contain 25000 viable seeds/m^2 in the plough layer and, although the average number is nearer to 5000/m^2, only a small proportion of even the 'cleanest' fields have less than 1000 seeds/m^2 (Roberts 1982). The species composition of the seed bank is influenced by previous cropping, cultivations and herbicide use, so that, in years when cruciferous crops are grown, a disproportionately high number of cruciferous weed seeds can be added to the soil. The evidence suggests that for a persistent seed bank in a cultivated soil, seed loss varies between 20% and 50% per year (Roberts 1982). The rate of loss is greatest for seeds near the soil surface and can be extremely slow for seeds, of cruciferous species like <u>Sinapis arvensis</u>, buried under grass crops.

Table 1. Mean numbers of cabbage root fly pupae recovered from the soil around the roots of cruciferous weeds during 1971 to 1973.

Plant species	Common name	*Mean number of pupae/10 plants		
		1971	1972	1973
<u>Sinapis arvensis</u>	Charlock	2	2	3
<u>Thlaspi arvense</u>	Field penny-cress	1	0	6
<u>Sisymbrium officinale</u>	Hedge mustard	9	0	5
<u>Capsella bursa-pastoris</u>	Shepherd's purse	0	0	0
<u>Raphanus raphanistrum</u>	Wild radish	25	20	28

*L.S.R. = 1.5

From the information presented, there appear to be insufficient wild-host plants in hedgerows and uncultivated areas to support cabbage root flies in sufficient numbers to add significantly to the overall pest population. When we first became interested in this subject in 1971, we tried to collect insects from wild and cultivated plants to determine whether the two were part of the same population. The numbers of wild-host plants found in isolated areas were too few to even initiate such a study. The only places we could find wild cruciferous plants in any numbers were in cultivated cruciferous crops and even in these only five species (Table 1) were present in numbers sufficient for meaningful sampling. Observations also indicated that flies preferred to lay initially on the cultivated plants and laid eggs on the weeds only when the cultivated plants started to mature.

During 3 years when wild-plants were sampled, most ($\underline{P}$ = 0.05) pupae were recovered from around the roots of <u>Raphanus raphanistrum</u> and no pupae were recovered from <u>Capsella bursa-pastoris</u>. Under the most favourable conditions in the field, the maximum numbers of cabbage root fly pupae recovered from wild-radish plants averaged 4.3/plant (Finch and Ackley 1977). Assuming a density of one wild-radish plant per metre around the edge of the cropped area, a considerable overestimate for most British situations, comparable wild-hosts would contribute a total of 1720 flies to a 1 ha brassica field, or 860 females, assuming the normal 1:1 sex ratio. If the area being cropped was increased, however, to square fields of 4, 9 and 16 ha, then the border effect would become less pronounced and the wild plants would contribute only 430, 287 and 215 females respectively to each hectare of crop. In a 4 ha field it is usual during the spring generation to recover about 100 eggs/plant from cauliflower crops. This number does not take into consideration fly eggs eaten by predatory ground beetles (Hughes and Salter 1959). In most early-summer cauliflower crops there are about 25,000 plants/ha and since each female lays between 50-60 eggs under field conditions (Finch 1971) each cultivated plant could on average be expected to receive one egg from the females from the wild-host plants. Thus insects developing on wild-host plants could contribute only about 1% of the root fly eggs laid on the cultivated plants under the conditions proposed assuming all females mated successfully and survived to lay. In addition to the local insects, others could arrive from wild host-plants from further afield. However, because of the herbicides used in the rotation, it is unusual for cruciferous weeds to survive in non-cruciferous crops. The main source of infestation from wild crucifers must therefore be other brassica fields. If, individuals from the wild-host plants can migrate such distances, so presumably can those from the cultivated host plants. It is also important to remember that the individual plants within cauliflower crops are relatively well spaced, compared to those in swede (<u>Brassica napus</u> var. <u>napobrassica</u>) crops where the target density is 160,000 plants/ha or in oilseed rape (<u>B. napus</u> var. <u>oleifera</u>) crops where it reaches 1,000,000 plants/ha. Obviously in crops where the plants are grown as densely as this, the influence of infestations from wild-hosts has to be negligible in that particular season.

MAJOR SOURCES OF INSECT PESTS OF CRUCIFEROUS CROPS

If, as suggested in the previous section, wild-host plants contribute little to the overall pest population within any one season, then what are the major sources of infestation? The results in Table 2 show clearly that cultivated plants are responsible for maintaining this pest. All of the cultivated plants, even those treated 'correctly' with insecticide, supported as many, or more, insects per plant than their untreated wild-host counterparts, with the exception of wild-radish.

Table 2.   Numbers of cabbage root fly pupae recovered from the soil around the roots of cultivated cruciferous plants in 1971 and 1973

Crop	Mean no. of pupae/(P)			
	Treated plants		Untreated plants	
	P/Plant	P/m^2	P/Plant	P/m^2
Brussels sprouts	1.0	2	7.9	18
	17.8*	36	19.0	43
Cabbage	0.9	5	6.3	38
Cauliflower				
Density (4/m^2)	0.7	3	13.5	54
Density (40/m^2)	1.0	40	2.8	110
From 7.5 cm pots	0.6	2	25.8	103
Swedes	2.0	32	8.6	138
Oilseed rape				
Heavy soils	0.1	6	1.6	96
Light soils	0.4	24	18.0	1080

* Diazinon treated; remaining crops treated with chlorfenvinphos

However, even when a highly-effective insecticide like chlorfenvinphos is incorporated 'correctly' into the soil, a proportion of the pest population always survives the treatment. Hence, because of their high numbers, plants treated with insecticide are a major source of pest infestation. The results in Table 2 show that 1-2 pupae were produced from each treated plant and that about 10 times as many were produced on 'untreated' plants. It is also evident from the diazinon-treated crop that if the chemical is not applied 'correctly', then for all intents and purposes the crop is untreated and will add large numbers of flies to subsequent generations.

Other treated crops that produce high pest infestations are those, such as swedes, that remain in the soil throughout the season. In such crops, the soil-insecticides applied at drilling control the spring generation of flies but do not persist sufficiently long to control the summer and autumn generations. Control of these generations is normally attempted by applying overhead sprays. Unfortunately this is not very sucessful, and, as a consequence, large numbers of cabbage root fly larvae usually develop on such crops.

The other major source of high pest infestation is the untreated plants. Plants established during the inter-generation periods will tolerate considerable damage, when eventually infested, before yield is affected. In such situations, particularly if damage is indirect, growers may not apply insecticide. In other instances growers transplant cauliflowers from 7.5 cm diameter pots, early in the season. Plants from such pots retain about 12 times as much root as comparable pulled plants and hence tolerate considerably more insects on their roots before damage becomes evident. The results in Table 2 show how plants of this type can be a major source of infestation.

WILD HOST PLANTS AS RESERVOIRS FOR PREDATORS OF CROP PESTS

Work at Wellesbourne in the late 1950's (Hughes and Salter 1959) and early 1960's (Coaker and Williams 1963) showed that predators exerted an important check on cabbage root fly populations, ground beetles alone killing more than 90% of the 200-300 eggs and first-instar larvae regularly found around the base of individual plants. In recent years, however, averages of 150 third-instar larvae/plant have occasionally been recorded from certain crops, indicating that the earlier estimates of predation may no longer be appropriate. The early estimates were obtained when high residues of aldrin and dieldrin, the commonest insecticides used at the time, were present in many soils. These persistent chemicals killed a wide range of species other than the target insects. Hence ground beetle predators surviving such hostile environments may have relied heavily upon food, such as cabbage root fly eggs, that arrived in the crop as a result of the immigration of pest insect species. With other sources of food, e.g. worms, in short supply, the ground beetles may have had little choice other than to feed voraciously upon cabbage root fly eggs. In contrast, it is possible that the use of the less-persistent organophosphorus insecticides over the last 20 years has allowed the systems to revert closer to the status they were in prior to the dieldrin era.

Even from the outset, however, workers in other countries disputed the English conclusions. Abu Yaman (1960) and van Dinther (1972), working independently in Holland, both concluded that mortality due to predators was small and according to Abu Yaman, affected only about 6% of the immature life stages of the pest. Similar sentiments were expressed by Mukerji (1971) in Canada, who estimated that about 17% of the pests' mortality was attributable to beetle predators.

In common with these workers, recent experiments at Wellesbourne have failed to demonstrate that predatory ground beetles now consistently kill a large percentage of the immature stages of the cabbage root fly (Finch and Skinner 1987). Our results demonstrated, however, that weather is an important limiting factor in the absence of which competition between larvae limits their numbers in a density-dependent way. Although at Wellesbourne there is still a 90% mortality during the life-cycle of the fly, only about 30% of this mortality now appears to occur at the egg and early larval stages.

The results therefore raise doubts about whether experiments to improve the effectiveness of predatory ground beetles will ever be successful. If abiotic factors are responsible for a high percentage of the overall mortality, there may be no advantage from releasing more efficient predators if they only kill those life-stages that would normally be killed by abiotic factors.

It is also important to remember that the ground beetles are largely opportunists and eat more or less any prey they can capture. This lack of specificity means they find suitable food in both cereal and brassica fields, and since many of the predators are apterous, it is unlikely that they would evolve any close associations with a particular group of phytophagous insects. Irrespective of which pest of cultivated horticultural field crops is being described, it is usual to see the same list of ground beetles mentioned for their general beneficial predatory qualities. Such beetles however, are common to the habitat and like the crucifers seem to prefer disturbed situations. The beetles, largely _Carabidae_ and _Staphylinidae_, are present in high numbers in most cultivated fields.

Owing to the relatively small areas of land surrounding crops from which additional beetles could be recruited, it seems unlikely that maintaining diversity in hedgerows will help to increase the overall numbers of predatory ground beetles. The hedgerow itself is more an extension of a woodland habitat, where the beetle species found differ from those found in the open field (Pollard 1968). In addition, Speight and Lawton (1976) showed that fewer of the carabid and staphylinid beetles that form the bulk of the natural predator complex, were caught alongside hedgerows than in the rest of the field. This observation runs counter to the popular view that hedgerows provide an important reservoir for predators and thus enhance the regulation of crop pests (Speight and Lawton 1976).

The strong preference for a particular habitat was also demonstrated by Smith (1976), in her study of how weedy and weed-free plots of Brussels sprouts affected pest and predator populations. She showed that high numbers of the predatory anthocorid, <u>Anthocoris nemorum</u>, congregated on weeds and as a result many more were found on Brussels sprouts surrounded by weeds than on those surrounded by bare soil. Bare ground between Brussels sprouts plants, however, deterred the anthocorids from moving onto the plants in the plots, even when the crop plants were heavily-infested with aphids, again indicating that factors other than a plentiful food supply restrict certain predators to the more stable habitat found alongside hedgerows. Hence high numbers of predators in hedgerows do not necessarily mean they will assist in pest control in adjacent crops.

A similar conclusion was arrived at by Pollard (1971) concerning the regulation of <u>Brevicoryne brassicae</u> populations by syrphids. He studied the influence of hedges not just surrounding a single field, but in two areas of sharply contrasting farmland. In one, hedgerows were numerous and in the other they were scarce. His results showed that the composition of the surrounding countryside had no effect on syrphid oviposition on aphid-infested Brussels sprouts. In fact more syrphid eggs were recorded on the infested plants in the areas without the hedgerows than in those with. In an earlier study, van Emden (1965) reported that syrphid oviposition on Brussels sprouts plants infested with <u>Brevicoryne brassicae</u> was heavier alongside a flowering hedgerow. However, in a similar situation, Chandler (1968) was unable to demonstrate that hedgerow flowers increased syrphid oviposition locally and suggested that van Emden's (1965) results arose merely from more syrphids concentrating in the area of the crop sheltered by the hedge. It is also doubtful whether there is anything to be gained from conserving flowering plants in hedgerows to provide nectar and pollen for predators, since the same sources of food

54

are utilized by the adults of many pest species (Finch 1971).

The conclusions from our studies of a soil pest and Pollard's (1968) of a foliage pest are similar. Both indicate that the contribution of hedgerows to beetle predator populations is insignificant and that diversity of habitat outside the crop does not add to the stability of crop pest populations.

WILD HOST PLANTS AS RESERVOIRS FOR PARASITOIDS OF CROP PESTS

So few cabbage root fly pupae are produced on wild plants outside brassica fields that the numbers of parasitoids arising from such sources must be negligible. In contrast, the numbers of parasitoids produced in cultivated crops can be prodigious. When an untreated swede crop was grown at Wellesbourne in 1975, about 70% of the cabbage root fly pupae recovered were parasitised by either the cynipid Idiomorpha rapae (Westw.), or the staphylinid Aleochara bilineata. In England levels of parasitism are not generally as high as this, largely because most crops are treated at some time with an insecticide and the parasitoids are more susceptible to the chemicals than the pest (Finch and Collier 1984). Idiomorpha rapae, which lays its eggs on the first- or second-instar larvae has been recorded from more than 60% of cabbage root fly pupae in Canada (Wishart et al. 1957). Aleochara bilineata and A. bipustulata (L.), whose larvae bore into newly-formed pupae, regularly parasitize 20-30% of the cabbage root fly pupae, and occasionally 60% (Wishart et al. 1957). There are, therefore, often large numbers of parasitoids emerging in the spring from the fields of the previous years crops. Both parasitoids appear to be synchronized with the pest population of the cultivated crops, since at 20°C Idiomorpha adults emerged 1 week and Aleochara adults 2 weeks after peak emergence of the flies (Finch and Collier 1984).

It has been suggested that the parasitoids of some of the foliage pests are found frequently on wild-host plants. However, van Emden (1965) showed that parasitized Brevicoryne brassicae were evenly distributed across experimental plots even in the area alongside the hedgerow. Similarly, Root (1973) showed that adult parasitoids were present in similar numbers per plant on both 'single rows' and 'pure stands' and that in the large samples taken towards the close of each season, the level of aphid parasitism was similar on both treatments. Root (1973) also stated that predators and parasitoids were always relatively rare on his test plants, especially during the early part of each season, indicating they were probably travelling some distance (the previous crop field) rather than just arriving from local wild-host plants.

Another distant traveller, <u>P. rapae</u>, appears to be subjected to considerable selection pressures from parasitoids, particularly in agricultural systems. Larvae placed on potted plants among grass or flower beds were never parasitized by the braconid <u>Apanteles rubecula</u>, whereas similar larvae placed on pots in cabbage plots were heavily parasitized (Richards 1940). Therefore, Richards concluded that the parasitoids were attracted first to large areas of cabbage and only subsequently proceeded to hunt for the Pierid larvae. Parasitism of this type may favour individuals which lay on isolated plants. It also indicates that as many wild host plants are relatively isolated they are less likely to be reservoirs of parasites of crop pests.

Adults of <u>Diaeretella rapae</u> were also attracted first to the vicinity of cultivated cruciferous plants, and only subsequently to their primary host, the cabbage aphid (Read et al. 1970). It is easy to suggest how this mechanism has evolved, since Root and Tahvanainen (1969) never found <u>Brevicoryne brassicae</u> infesting even the commonest wild crucifer, <u>Barbarea vulgaris</u>, in their region of New York State. Root and Olson (1969) also showed that when plants of <u>Barbarea vulgaris</u> and three cultivars of <u>Brassica</u> were infested with colonies of <u>Brevicoryne brassicae</u>, <u>Barbarea vulgaris</u> was the most resistant host. It was also the only host on which the original colonies soon produced alates, a general indicator of a non-preferred plant.

WILD HOST PLANTS AS RESERVOIRS OF PEST GENOTYPES

During the elaborate co-evolution between plants and phytophagous insects, many plants have evolved a variety of natural pesticides, only a few of which can be metabolized by any given insect species. Monophagous insects conserve energy by restricting their countermeasures to one or a few potentially toxic substances (Brower et al. 1958). In contrast, polyphagous species have to counter a wider range of potential toxins, many of which are present in high concentration in certain food plants.

According to Krieger et al. (1971) the activity of midgut microsomal oxidase enzymes is higher in polyphagous than in monophagous species because the natural function of these enzymes is to detoxify natural insecticides present in the larval food plants. These enzyme system are also effective against a wide range of synthetic insecticides and hence are involved in detoxifying chemicals such as dieldrin (Krieger et al. 1971) and pyrethrins (Doskotch and El-Feraly 1969). If, therefore, populations that regularly develop on different wild-host plants have a greater propensity for dealing with defensive compounds, they may be more likely to develop resistance to our current insecticides. Alternative host-plants could be detrimental, therefore, by adding greater heterogenity to the pest population. Whether or not this occurs will depend largely upon whether the few individuals from wild host plants are truly oligophagous, and move freely from the wild to the cultivated host plants, or whether they also restrict themselves to a relatively narrow host range by remaining on the wild plants.

HOW HAS CABBAGE ROOT FLY DEVELOPED PEST STATUS?

The cabbage root fly is restricted to the temperate zone of the holarctic region (35-60°N) (de Wilde 1947). It is generally agreed that, throughout the whole of this zone, flies from the overwintering population emerge in April and May and that, once started, the whole population emerges over a relatively short period of time (see Coaker and Finch 1971). In a recent survey, however, Finch and Collier (1983) showed that emergence from overwintering puparia in certain localities in England and Wales was protracted, many flies not emerging until June or July. Similar delays in emergence were also recorded from several areas in Europe (Finch et al. 1987 - In Press). Attempts to determine which factors regulated this 'late-emergence' (Collier and Finch, 1987 - In Press) revealed that 'lateness' was controlled by a single dominant gene. Hence, cabbage root fly was probably restricted originally to just two generations per year. Under such conditions there would be little necessity for the fly to breed on wild crucifers in the spring and autumn to maintain its population from one year to the next. As wild crucifers have co-evolved along with the insects, it now seems more likely, that the present three-generation system has evolved in response to agricultural changes such as extending the season for cultivated crops and growing highly-nutritive cultivars that enable the pest to complete its life-cycle in a shorter period.

INTERSPECIFIC COMPETITION BETWEEN INSECTS

Few studies have been made on intra specific behavioural relationships during host-plant selection (Rothschild and Schoonhoven 1977) and even fewer on interspecific relationships. This aspect of host plant selection was mentioned by Root (1973) in his comment that "crucifer aphids tend to move off experimental plants that are moderately damaged by P. cruciferae (D.P. Read and J.J. Skelsey pers. comm.)" but until recently (Finch and Jones, 1987) interspecific competition has not been studied in detail.

Host-plant selection by the cabbage root fly is now known to be influenced considerably by other insects competing for the same resource. Traynier (1967) showed that although cabbage root fly females lay in the soil adjacent to plant stems, they first have to probe the surface of the plant to determine whether it is a suitable site for oviposition. During this process, host-plant selection can be influenced by the frass of other insects and also by the plant type the original colonizers are feeding on (Finch and Jones 1987, cf. Renwick and Radke 1985). Species of white butterflies (Pieris spp.) positively advertised their presence to cabbage root fly by producing larger amounts of chemicals similar to those released normally from damaged plants. Although such species reduced the sulphur compounds in their diet to nitriles in their frass, the changes did not deter cabbage root fly from ovipositing. In contrast, larvae of the mustard beetle (Phaedon cochleariae) and the cabbage moth (Mamestra brassicae) seemed able to disguise their activity on four of the eight host plants tested and produced frass that did not influence host-plant selection by cabbage root fly. This was not the case with the garden-pebble moth (Evergestis forficalis); irrespective of host-plant, all frass produced by its larvae deterred cabbage root fly oviposition. Cabbage root fly oviposition was also deterred when the diamond-back moth (Plutella xylostella) had previously laid on the plants, indicating that a marking pheromone from this moth may have influenced cabbage root fly behaviour. Finally, cabbage root fly was deterred from laying on plants infested by the cabbage aphid (<u>Brevicoryne brassicae</u>) by the physical disturbance of the aphids or an alarm pheromone. Infested plants were more preferred when the actual aphids, but not their honeydew, were removed immediately prior to testing (Finch and Jones 1987).

Therefore, host-plant selection can be influenced considerably by other insects infesting the same host-plant, and not just by those attempting to occupy the same niche.

INFLUENCE OF ECOSYSTEM TEXTURE

According to the Resource Concentration Hypothesis of Root (1973), phytophagous insects are more likely to find and remain on host plants that occur in pure and/or dense patches. Such a tendency is most pronounced among specialist feeders whose life requirements are closely tuned to conditions within a simple stand of a particular host plant.

When plant concentration was altered by planting pure collard stands of different sizes (1, 10 or 100 plants), however, the predictions of the Resource Concentration Hypothesis were only partially fulfilled (Cromartie 1975). Some specialist insects such as the collard flea beetle, <u>P. cruciferae</u>, did achieve their greatest densities in large plots, but others such as the cabbage butterfly <u>Pieris rapae</u> (L.) preferred the isolated single plants. According to Root (1973), the results demonstrated that phytophagous insects respond individually to textural properties of vegetation such as patch size. The main factor that seems to be overriding, however, is that the adults actively seek their own preferred habitat.

The observation that more <u>P. rapae</u> eggs were found on isolated plants can also be explained by the behaviour of the females (Jones 1977). Ovipositing females turned back on leaving areas containing host plants and hence laid more eggs on plants at the edges of plots and on isolated plants.

CAN PEST INFESTATIONS BE REDUCED BY INCREASING WITHIN-CROP DIVERSITY?

In 1969, Dempster compared the effects of different types of weed control on the numbers of the small-white butterfly (<u>Pieris rapae</u>) infesting Brussels sprouts plants.

He concluded that survival of caterpillars was lower in weedy crops but that the advantage gained from such a 'diverse' system had to be balanced against the harmful effect of competition between the weeds and the crop. He suggested that the benefits from weed-cover could possibly be obtained by undersowing brassica crops with clover (<u>Trifolium</u>), a plant that would provide a habitat suitable for predatory beetles but not compete unduly with the crop. When this was attempted by completely undersowing plots of Brussels sprouts (Dempster and Coaker 1974), although populations of the small-white butterfly, the cabbage aphid and the cabbage root fly were lowered, the competition from the clover was still too severe. To overcome this difficulty, O'Donnell and Coaker (1975) grew strips of recumbent clover (cv. Kersey White) between rows of Brussels sprouts plants so that only 25, 50 and 75% of the ground was covered. Using a clover crop in this way reduced oviposition by cabbage root fly through the presence, the authors suggested, of too much vegetation. When the densities of cabbage, cauliflowers, Brussels sprouts, and swede, were changed sequentially from $1.5/m^2$ to $83/m^2$ in a systematically designed experiment, however, there was no reduction in the numbers of cabbage root fly eggs laid after the canopy of the adjacent rows of plants became continuous. The pattern of oviposition was not affected by dense stands of vegetation when the vegetation consisted of host plants, provided that there was bare soil beneath the plants (Finch and Skinner 1976).

This bare-soil effect was stressed by Smith (1976) who showed that clean-weeding of Brussels sprouts crops provided ideal conditions for colonization by aphids, whitefly and certain Lepidoptera. The preference by certain pest insects for plants that stand out against a background of bare soil (Root & Tahvanainen 1969) may be partly due to the insects' phototactic reaction and partly to the optomotor reaction resulting from plants 'looming up' along the insects' flight path (Kennedy et al. 1961). This optomotor effect is obviously accentuated when plants are highlighted against a bare-soil background, a characteristic of most 'spaced' arable crops. In effect, therefore, the weedy background may act to camouflage the plants' silhouette so that it merges more successfully with its background.

The field experiments of Smith (1976) also showed that a background of weeds decreased populations of the cabbage aphid on Brussels sprout plants. No evidence was presented, however, to indicate whether this was due to the different levels of immigration initially, or to a different rate of development of the colonies once the aphids had established. Smith (1976) stressed, however, that the plants growing in the weedy plots suffered considerably from competition. It seems equally likely, therefore, that the differences in aphid numbers arose largely as a result of the better plant growth in the weed-free plots enabling the aphid colonies to increase at a faster rate.

Although most authors (e.g. Pimentel 1961, Dempster 1969, Smith 1969, Root 1973, Dempster and Coaker 1974, O'Donnell and Coaker 1975, Smith 1976) who grew crop plants in weedy or diverse backgrounds, indicated that the diverse situations increased the levels of predation on their test plants, the evidence to support such a conclusion is extremely scanty. For example, the suggestion of Dempster and Coaker (1974) that predation of caterpillars of the small-white butterfly was higher on Brussels sprouts undersown with clover was based on the premise that similar numbers of eggs were laid on the plants in both their weedy and the weed-free plots. However, the authors freely acknowledged that their original premise was based largely on heavy oviposition on just two plants in one treatment. Hence, had they taken larger samples, they may have had to conclude, as they did for the cabbage aphid and the cabbage root fly, that any changes in survival on plants growing under the different cultural regimes "could not be separated from those of immigration" (Dempster and Coaker 1974). As stated in detail earlier, many of the other low insect infestations on plants growing in diverse backgrounds are just as easy to explain by reduced plant growth as by increased predator activity. Support for this can be found in many papers, such as the one of Tahvanainen and Root (1972) where they concluded that "predators and parasites appeared to have negligible influence on adults of the collard flea beetle (<u>Phyllotreta</u> <u>cruciferae</u>) infesting collards growing either in monocultures or adjacent to natural vegetation". Similarly Theunissen and Den Ouden (1980), who intercropped Brussels sprouts with the weed Corn Spurrey (<u>Spergula</u> <u>arvensis</u>), showed that the reduction in number of cabbage aphid, cabbage moth, garden-pebble moth and cabbage root fly all resulted from the weeds interfering during host-plant selection by the adult insects and not from enhanced predation.

An alternative to growing weedy crops, is to grow two crop plants together. The easiest system is to grow both crops in rows (one form of 'row intercropping').

Tahvanainen and Root (1972) adopted this approach, and demonstrated that monocultures of collards were colonized more rapidly by the collard flea beetle, and experienced greater feeding damage, than stands inter-planted with tomatoes and tobacco. Continuing this approach, Buranday and Raros (1973) suggested that a 50% ground cover was most effective in lowering diamond-back moth oviposition on cabbage when tomatoes were used as the intercrop.

Intercropping brassicas with French beans (Phaseolus vulgaris) and grass (Lolium spp) also reduced infestations of the cabbage aphid and the cabbage root fly by over 60% compared with those on brassicas grown in pure stands (Tukahirwa and Coaker 1982). Twice as many carabid and staphylinid predators of the immature stages of rootflies were trapped on intercropped as on pure brassica stands. However, similar reductions in root fly eggs occurred when predators were excluded, suggesting that predation was not an important factor suppressing root flies in intercrops.

Results from both laboratory and field tests suggest that the increased locomotor activity of the flies in the mixed plots led to their more rapid departure from the mixed plots than from the brassica plots and therefore correspondingly fewer eggs were laid on the mixed plots (Tukahirwa and Coaker 1982). Thus the disturbance of oviposition behaviour in mixed stands reduced the exploitation of the brassicas by the flies, a situation comparable with the limited infestation of wild host plants (Finch and Ackley 1977). The results of Tukahirwa and Coaker (1982) suggested that a maximum reduction in cabbage aphid and cabbage root fly attack in intercropped brassicas is obtained when the rows of the two plant types are 50 cm apart of less. The diversionary effect of the non-host plant was then at its greatest and the maximum reduction occurred in pest oviposition.

Single-row intercropping, therefore, appears to be the best arrangement of plants for reducing pest attacks. It has been most effective when the intercrop provided at least 50% ground cover between the rows at the time of pest invasion (Buranday and Raros 1973, Theunissen and Den Ouden 1980, Tukahirwa and Coaker 1982). Perrin and Phillips (1978) showed, however, that similar numbers of eggs/plant were laid by the diamond-back moth and cabbage whitefly (Aleyrodes proletella) in a monocrop of Brussels sprouts as in an alternate row inter-crop in which the density of Brussels sprouts was halved. Therefore, the advantage is gained from this type of intercropping only if the host-plant density is maintained the same as in the corresponding monoculture (Perrin and Phillips 1978). In addition, changes due to intercropping may not all be beneficial. Where one or two specific natural enemies are more important than a whole complex of generalized predators, intercropping may result in harmful interference and disruption of the most important natural enemies (Perrin and Phillips 1978). Finally, the choice of the plant used as the intercrop is extremely important. Lettuce can be too competitive with cabbage and decrease the subsequent cabbage yield by 60% (Ryan, Ryan and McNaedhe 1980).

As to be expected, the 'diversionary effect' could work in several different ways. According to Tukahirwa and Coaker (1982), similar numbers of cabbage root fly entered plots of cabbage intercropped with beans or grass as entered pure stands, indicating that non-host plants did not interfere with this fly's oriented response to its host-plants. In such situations however, there is also a temporal factor that needs to be considered. When colonizing insects arrive shortly after the crop and its intercrop have been planted, there is unlikely, as Tukahirwa and Coaker (1982) conclude, to be any major effect on pest infestation as a consequence of the intercrop. However, this situation may not prevail throughout the growth of the crop since, as soon as the two crops start to compete, the pest populations will also be affected. Obviously, the later that colonizers arrive in the season, the greater the effect the intercrop is likely to have on their initial colonization.

There is also some experimental evidence that odours from the intercropped plant can deter or repel certain pests. In choice experiments in the laboratory, Tahvanainen and Root (1972) showed that chemical stimuli given off from tomato (<u>Lycopersicon</u> <u>esculentum</u>) and ragweed (<u>Ambrosia</u> <u>artemisiifolia</u>) extracts interfered with host-plant finding by the collard flea beetle. Similarly, Buranday and Raros (1973) suggested that the odours from tomatoes intercropped with cabbage were responsible for lowering oviposition by the diamond-back moth on the cabbage plants. However, no evidence was provided to show that the moths were being repelled, so this may just be a 'diversionary effect' operating in a manner similar to that described by Tukahirwa and Coaker (1982). This seems the more likely explanation in this case, since Perrin and Phillips (1978) showed that the odour of tomatoes failed to reduce damage by the diamond-back moth, on Brussels sprouts, when tomatoes and Brussels sprouts were intercropped in glasshouse experiments. Many claims are also made that African marigolds (<u>Tagetes</u> spp.) planted between rows of crop plants reduce pest numbers. Whether this is achieved through a direct effect of the odours of the African marigold repelling the colonizing insects has not been elucidated. It is well known that species of African marigolds release large amounts of root exudates which can be taken up by adjacent plants (Rovira 1969). It is possible, therefore, that a crop interplanted with African marigolds is affected directly as a result of root uptake rather than by having its odour masked.

CONCLUSIONS

The one major factor that has emerged during the compilation of this review has been the overriding influence of habitat. In general, the cruciferous plants selected as crop plants are those found growing naturally in disturbed ground and the insect species that have developed pest status on such plants are also those that prefer their host-plants well-spaced and clearly visible. Attempts to lower pest infestations by diversifying the habitat outside the crop have largely been unsuccessful.

The fear that wild-host plants contribute large numbers of insects to an overall pest population in the short term appears to be unfounded. It should be remembered, however, that owing to the effectiveness of modern herbicides, weeds are now much less abundant, and hence much less of a problem than they were when the advice to destroy all cruciferous weeds was being advocated (Ullyett 1947). In the present situation, the contribution to the overall pest population of cruciferous crops from wild host-plants is negligible, since there are few areas where wild host-plants are present in sufficiently high numbers to produce an effect. Also, where they do occur in reasonable numbers, mainly in brassica crops, they are usually less preferred than the cultivated plants. The only possible threat from pest insects developing on wild cruciferous plants is that they could add heterogeneity to the population and hence increase its potential for developing resistance to some of our existing insecticides. The likelihood of this occurring, however, could be more apparent than real, since most insects developing on weeds seem to be part of the population that normally develops on the cultivated plants, rather than vice versa.

64

There have also been numerous attempts in the past to increase the diversity of plants within hedgerows and other uncultivated areas in the hope that this would enrich the populations of natural enemies and lead to better biological control in neighbouring cultivated fields. To increase the numbers of parasitoids and predators specific to cruciferous pests, however, cruciferous plants attractive to crop pests first have to be established in hedgerows. Unfortunately, most of these plants do not compete well in the hedgerow environment and are soon smothered by the more competitive hedgerow plants. Although increasing the heterogeneity in hedgerows undoubtedly increases the numbers of parasitoid and predatory insects found there, the changes are most probably due largely to increases in the species restricted to hedgerows, rather than to increases in the numbers of those species normally associated with cultivated crops. For the cabbage root fly, and for many of the other major pests of cruciferous crops discussed in this review, wild host-plants are not a large reserviour for pest species, nor do they add sufficient numbers of parasitoids to assist significantly in the day-to-day pest control in cultivated crops. Usually the pests and their associated parasitoids move from one host crop to the next and although they are sometimes accompanied by a few specific predators, such as syrphids and coccinellids, a large number of the major predators are already in the field where they feed as opportunists on any insects that arrive to colonize whatever crop is planted in the field.

The most obvious alternative to reducing pest numbers by increasing diversity alongside the crop would seem to be by increasing diversity within the crop. Surprisingly, this also has not produced any clear-cut benefits. Admittedly, intercropping does reduce pest numbers, but the mechanism involved is still somewhat unclear. If as suggested by certain workers (Tahvanainen and Root 1972, Buranday and Raros 1973) the intercropped plants repel the colonizing insects directly by odour, or by indirect effects mediated through the cruciferous host plant, then this approach may in future prove promising. At present, however, there is a considerable danger in drawing any definite conclusions concerning the effectiveness of intercropping, since all systems tested to date have affected the growth of the host plant, which in itself is sufficient to reduce pest numbers.

The whole problem of pest control in cruciferous crops arises from the fact that by providing cultivated cruciferous plants with near-to-ideal growing conditions, and by selecting rapidly maturing cultivars, the specialist insects are not only provided with an abundant source of food but also often with a highly-nutritive one. It is not surprising, therefore, that certain insects have developed pest status once away from the constraints that abound in diverse natural communities. By the same token, since the latter communities support lower pest populations and are more stable than monocultures, they are unlikely to harbour a surfeit of beneficial insects. Hence, it is probably unrealistic to expect insects from diverse stable systems, to contribute significantly in reducing pest populations in cultivated crops even if we could induce them to leave their more preferred stable habitats. Other ways will have to be found.

In the past, although spatial rotation has been practiced, it is usually only a matter of time before the pest insects locate the new host crops. Hence, using temporal rotation to provide a break in the cropping regime will probably prove more rewarding, provided it can be implemented and the pest species is incapable of migrating in from neighbouring areas or countries where such a rotation is not employed.

REFERENCES CITED

Abu Yaman, I.K. (1960). Natural control in cabbage root fly populations and influence of chemicals. *Meded. LandbHoogesch. Wageningen* 60, 1-57.

Andrewartha, H.G. and Birch, L.C. (1954). *The distribution and abundance of animals*. University of Chicago Press, Chicago.

Brower, L.P. (1958). Bird predation and foodplant specificity in closely related procryptic insects. *Am. Nat.* 92, 183-187.

Buranday, R.P. and Raros, R.S. (1973). Effects of cabbage-tomato intercropping on the incidence of oviposition on the diamond-back moth, *Plutella xylostella* (L.). *Philipp. Ent.* 2, 369-374.

Chandler, A.E.F. (1968). Some host-plant factors affecting oviposition by aphidophagous Syrphidae (Diptera). *Ann. Appl. Biol.* 61, 425-446.

Chong, C. and Bible, B. (1974). Variation in thiocyanate content of radish plants during ontogeny. *J. Amer. Soc. Hort. Sci.* 99, 159-162.

Clapham, A.R., Tutin, T.G. and Warburg, E.F. (1958). *Flora of the British Isles*. London: University Press.

Coaker T.H. and Finch, S. (1971). The cabbage root fly *Erioischia brassicae* (Bouche). *Rep. Natn. Veg. Res. Stn. 1970*, pp. 23-42.

Coaker, T.H. and Williams, D.A. (1963). The importance of some Carabidae and Staphylinidae as predators of the cabbage root fly, *Erioischia brassicae* (Bouché). *Ent. Exp. & Appl.* 6, 156-164.

Cole, R.A. (1980). Volatile components produced during ontogeny of some cultivated crucifers. *J. Sci. Fd & Agric.* 31, 549-557.

Collier, Rosemary H. and Finch S. (1987). Preliminary studies of the factors regulating late emergence in certain cabbage root fly, *Delia radicum*, populations. *IOBC/WPRS Bulletin 1986* (in press).

Cromartie, W.J. (1975). The effect of stand size and vegetational background on the colonization of cruciferous plants by herbivorous insects. *J. Appl. Ecol.* 38, 259-281.

Czapek, F. (1922). *Biochemie der Pflanzen*. Vols I-III. G. Fisher, Jena.

Dempster, J.P. (1969). Some effects of weed control on the numbers of the small cabbage white (*Pieris rapae* L.) on Brussels sprouts. *J. Appl. Ecol.* 6, 339-345.

Dempster, J.P. and Coaker, T.H. (1974). Diversification of crop ecosystems as a means of controlling pests. In: D. Price Jones & M.E. Solomon (Eds.): *Biology in Pest and Disease Control*. Blackwell, Oxford. 106-114.

Dethier, V.G. (1970). Chemical interactions between plants and insects. In: *Chemical Ecology* (Sondheimer, E. and Simeone, J.B., eds), 83-102. Academic Press, New York and London.

De Wilde, J. (1947). De koolvlieg en zijn bestrijding. *Meded TuinbVoorlDienst* 45, 70.

Dinther, J. van (1972). Carabids as predators of the cabbage root fly. *Entomologia* (Ber.) 32, 193-194.

Doskotch, R.W. and El-Feraly, F.S. (1969). Isolation and characterization of (+)-sessamin and B-cyclopyretrosin from pyrethrum flowers. *Can. J. Chem.* 47, 1139-1142.

Ehrlich, P.R. and Raven, P.H. (1964). Butterflies and plants: a study in coevolution. *Evolution* 18, 586-608.

Ellis, P.R., Cole, Rosemary, A., Crisp, P. and Hardman, J.A. (1980). The relationship between cabbage root fly egg laying and volatile hydrolysis products of radish. *Ann. Appl. Biol.* 95, 283-289.

Elton, C.S. (1958). The Ecology of Invasion by Animals and Plants. London: Methuen.

Emden, H.F. Van (1965). The effect of uncultivated land on the distribution of cabbage aphid (Brevicoryne brassicae) on an adjacent crop. J. Appl. Ecol. 2, 171-195.

Feeny, P. (1976). Plant apparency and chemical defense. In: Biochemical interaction between plants and insects. Rec. Adv. Phytochem. 10, 1-40.

Finch, S. (1971). The fecundity of the cabbage root fly Erioischia brassicae (Bouché) under field conditions. Ent. Exp. & Appl. 14, 147-160.

Finch, S. (1977). Effect of secondary plant substances on host-plant selection by the cabbage root fly. In: Comportement des Insectes et Milieu Trophique, 251-268. Colloques Internationaux du C.N.R.S. No 265. C.N.R.S., Paris.

Finch, S. (1980). Chemical attraction of plant-feeding insects to plants. (Ed Coaker, T.H. London: Academic Press) Applied Biology 5, 67-143.

Finch, S. and Ackley, C.M. (1977). Cultivated and wild host plants supporting populations of the cabbage root fly. Ann. Appl. Biol. 85, 13-22.

Finch, S. and Collier, R.H. (1983). Emergence of flies from overwintering populations of cabbage root fly pupae. Ecol. Entomol. 8, 29-36.

Finch, S. and Collier, R.H. (1984). Parasitism of overwintering pupae of cabbage root fly, Delia radicum (L.) (Diptera: Anthomyiidae), in England and Wales. Bull. Ent. Res. 74, 79-86.

Finch, S. and Gairn, C.M. (1976). Mustard beetle. Rep. Natn. Veg. Stn. 1975, p. 84-85.

Finch, S. and Jones, T.H. (1987). Interspecific competition during host-plant selection by insect pests of cruciferous crops. Proc. of the 6th Int. Symp. on Insect Plant Relationships. Pau, France, 1-4 July 1986.

Finch, S. and Skinner, G. (1976). The effect of plant density on populations of the cabbage root fly (Erioischia brassicae (Bouché)) and the cabbage stem weevil (Ceuthorrhynchus quadridens (Panz.)) on cauliflowers. Bull. Ent. Res. 66, 113-123.

Finch, S. and Skinner, G. (1987). Mortality of the immature stages of the cabbage root fly. IOBC/WPRS Bulletin 1986 (in press).

Finch, S., Cole, R.A. and Skinner, G. (1975). Chemicals influencing cabbage root fly behaviour: Isolation of attractants. Rep. Natn. Veg. Res. Stn. for 1974, p. 96-97.

Finch, S., Bromand, B., Brunel, E., Bues, M., Collier, R.H., Dunne, R., Foster, G., Freuler, J., Hommes, M., Van Keymeulen, M., Mowat, D.H., Pelerents, C., Skinner, G., Stadler, E. and Theunissen, J. (1986). Emergence of cabbage root fly from puparia collected throughout northern Europe. IOBC/WPRS Bulletin 1987 (in press).

Freeman, G.G. and Mossadeghi, N. (1972). Influence of sulphate nutrition on flavour components of three cruciferous plants: radish (Raphanus sativus), cabbage (Brassica oleracea capitata) and white mustard (Sinapis alba). J. Sci. Fd & Agric. 23, 387-402.

Hardman, J.A. and Ellis, P.E. (1978). Host plant factor influencing the susceptibility of cruciferous crops to cabbage root fly attack. Ent. Exp. & Appl. 24, 393-397.

Hicks, K.L. (1974). Mustard oil glucosides: feeding stimulants for adult cabbage flea beetles, Phyllotreta cruciferae (Coleoptera: Chrysomelidae). Ann. Ent. Soc. Am. 67, 261-264.

Hughes, R.D. and Salter, D.D. (1959). Natural mortality of Erioischia brassicae (Bouché) (Dipt., Anthomyiidae) during the immature stages of the first generation. J. Anim. Ecol. 28, 231-241.

Jones, R. E. (1977). Search behaviour: a study of three caterpillar species. Behaviour, LX, 237-259.

Joseffson, E. (1967). Distribution of thioglucosides in different parts of Brassica plants. Phytochem. 6, 1617-1627.

Kennedy, J.S., Booth, C.O. and Kershaw, W.J.S. (1959). Host finding by aphids in the field. II. Aphis fabae Scop. (gynoparae) and Brevicoryne brassicae L.; with a re-appraisal of the role of host-finding behaviour in virus spread. Ann. Appl. Biol. 47, 424-444.

Kennedy, J.S., Booth, C.O. and Kershaw, W.J.S. (1961). Host finding by aphids in the field. III. Visual attraction. Ann. Appl. Biol. 49, 1-21.

Kjaer, A. (1963). The distribution of sulphur compounds. In: Chemical Plant Taxonomy (Swain, T., ed), 453-473. Academic Press, New York and London.

Krieger, R.I., Feeny, P.P. and Wilkinson, C.F. (1971). Detoxication enyzmes in the guts of caterpillars: An evolutionary answer to plant defenses? Science 172, 579-581.

Mukerji, M.K. (1971). Major factors in survival of the immature stages of Hylemya brassicae (Diptera: Anthomyiidae) on cabbage. Can. Ent. 103, 717-728.

O'Donnell, M.S. and Coaker, T.H. (1975). Potential of intra-crop diversity for the control of brassica pests. Proc. 8th Brit. Insect Fung. Conf. 1, 101-107.

Perrin, R.M. and Phillips, M.L. (1978). Some effects of mixed cropping on the population dynamics of insect pests. Ent. Exp. & Appl. 24, 585-593.

Pimentel, D. (1961a). The influence of plant spatial patterns on insect populations. Ann. Ent. Soc. Am. 54, 61-69.

Pimentel, D. (1961b). Species diversity and insect populations outbreaks. Ann. Ent. Soc. Am. 54, 76-86.

Pollard, E. (1968). Hedges. III. The effect of removal of the bottom flora of a hawthorn hedge on the Carabidae of the hedge bottom. J. Appl. Ecol. 5, 125-139.

Pollard, E. (1971). Hedges. VI. Habitiat diversity and crop pests: a study of Brevicoryne brassicae and its syrphid predators. J. Appl. Ecol. 8, 751-780.

Price, J.L. (1975). Diamond-back moth. Rep. Natn. Veg. Res. Stn for 1974, p. 97-98.

Prokopy, R.J., Collier, R.H. and Finch, S. (1983). Leaf colour used by cabbage root flies to distinguish among host plants. Science 221, 190-192.

Radcliffe, E.B. and Chapman, R.K. (1966). Varietal resistance to insect attack in various cruciferous crops. J. Econ. Ent. 59, 120-125.

Read, D.P., Feeny, P. and Root, R.B. (1970). Habitat selection by the aphid parasite Diaeretiella rapae (Hymenoptera: Braconidae) and hyperparasite Charips brassicae (Hymenoptera: Cynipidae). Can. Ent. 120, 1567-1578.

Renwick, J.A.A. and Radke, C.D. (1985). Constituents of host- and non-host plants deterring oviposition by the cabbage butterfly, Pieris rapae. Ent. Exp. & Appl. 39, 21-26.

Richards, O.W. (1940). The biology of the small white butterfly (Pieris rapae) with special reference to the factors controlling its abundance. J. Anim. Ecol. 9, 243-288.

Roberts, H.A. (1982). <u>Weed Control Handbook: Principles</u>. 7th Edition Oxford: Blackwell Scientific Publications.

Root, R.B. (1973). Organization of a plant-arthropod association in simple and diverse habitats: the fauna of collards (<u>Brassica oleracea</u>). <u>Ecol. Monogr.</u> 43, 95-120.

Root, R.B. and Olson, A.M. (1969). Population increase of the cabbage aphid, <u>Brevicoryne brassicae</u>, on different host plants. <u>Can. Ent.</u> 101, 768-773.

Root, R.B. and Tahvanainen, J.O. (1969). Role of winter cress, <u>Barbarea vulgaris</u>, as a temporal host in the seasonal development of the crucifer fauna. <u>Ann. Ent. Soc. Am.</u> 62, 852-855.

Rothschild, M. and Schoonhoven, L.M. (1977). Assessment of egg load by <u>Pieris brassicae</u> (Lepidoptera: Pieridae). <u>Nature, London</u> 266, 352-355.

Rovira, A.D. (1969). Plant root exudates. <u>Bot. Rev.</u> 35, 35-39.

Ryan, J., Ryan, M.F. and F. McNaeidhe (1980). The effect of interrow plant cover on poplations of the cabbage root fly <u>Delia brassicae</u> (Wiedemann). <u>J. Appl. Ecol.</u> 17, 31-40.

Schraudolf, H. and Bergman, F. (1965). Der Stoffwechsel von Indolderivaten in <u>Sinapis alba</u> L. II. Undersuchungen zur Biogenese und Umzetsung von Indolglucosinolaten mit Hilfe von Ringmarkiertem C^{14}-Tryptophan and S^{35}-Sulfat. <u>Planta</u> 67, 75-95.

Smith, J.G. (1969). Some effects of crop background in population of aphids and their natural enemies on Brussels sprouts. <u>Ann. Appl. Biol.</u> 63, 326-330.

Smith, J.G. (1976). Influence of crop background on aphids and other phytophagous insects on Brussels sprouts. <u>Ann. Appl. Biol.</u> 83, 1-13.

Southwood, T.R.E. (1962). Migration of terrestrial arthropods in relation to habitat. <u>Biol. Rev.</u> 37, 121-214.

Speight, M.R. and Lawton, J.H. (1976). The influence of weed-cover on the mortality imposed on artificial prey by predatory ground beetles in cereal fields. <u>Oecologia (Berl.)</u> 23, 211-223.

Tahvanainen, J.O. (1972). Phenlogy and microhabital selection of some flea beetles (Coleoptera: Chrysomelidae) on wild and cultivated crucifers in central New York. <u>Ent. Scand.</u> 3, 120-138.

Tahvanainen, J.O. & Root, R.B. (1972). The influence of vegetational diversity on the population ecology of a specialised herbivore, *Phyllotreta crucifera* (Coleoptera: Chrysomelidae). *Oecologia* 10, 321-346.

Theunissen, J. and Den Ouden, H. (1980). Effects of intercropping with *Spergula arvensis* on pests of Brussels sprouts. *Ent. Exp. & Appl.* 27, 260-268.

Traynier, R.M.M. (1967). Stimulation of oviposition by the cabbage root fly *Erioischia brassicae*. *Ent. Exp. & Appl.* 10, 401-412.

Tukahirwa, E.M. and Coaker, T.H. (1982). Effect of mixed cropping on some insect pests of brassicas: reduced *Brevicoryne brassicae* infestations and influence of epigeal predators and the disturbance of oviposition behaviour in *Delia brassicae*. *Ent. Exp. & Appl.* 32, 129-140.

Ullyett, G.C. (1947). Mortality factors in populations of *Plutella maculipennis* Curtis (Tineidae: Lep.), and their relation to the problem of control. Union of South Africa: Dept of Agric. and Forest., *Ent. Mem.* 2, 77-202.

Verschaffelt, E. (1910). The cause determining the selection of food in some herbivorous insects. *Proc. Acad. Sci. Amsterdam* 13, 536-542.

Wellington, W.G. (1957). Individual differences as a factor in population dynamics; the development of a problem. *Can. J. Zool.* 35, 293-323.

Whittaker, R.H. and Feeny, P. (1971). Allelochemics: Chemical interactions between species. *Science* 171, 757-770.

Wishart, G., Colhoun, E.H. and Monteith, A.E. (1957). Parasites of *Hylemya* spp. (Diptera: Anthomyiidae) that attack cruciferous crops in Europe. *Can. Ent.* 89, 510-517.

3

Entomology and Horticulture
of Muscadine Grapes

J. D. Dutcher, K. C. McGiffen, and J. N. All

INTRODUCTION

Grapevines are remarkable plants with an extensive capability for reproductive and vegetative growth (Pratt 1971, 1974). A complex system of long-lived axillary buds ensures their survival following injury or pruning. The stem stores and rapidly transports large amounts of water and nutrients. The whole plant matures in two or three years, supporting vigorous growth and berry production. The berries are a high sugar, high value crop that has been cultivated for 6000 years.

The grape genus <u>Vitis</u> is divided into 2 subgenera, <u>Vitis</u> and <u>Muscadinia</u>. Species of subg. <u>Vitis</u> have 19 pairs of chromosomes and those of subg. <u>Muscadinia</u> have 20 pairs. There are 29 species of subg. <u>Vitis</u> and 3 species of subg. <u>Muscadinia</u> native to North America (Fig. 1). Those in the subg. <u>Vitis</u> are important as 'bunch grapes' and as pest resistant rootstocks for <u>V</u>. <u>vinifera</u> L. The <u>Muscadinia</u> species, <u>V</u>. <u>rotundifolia</u> Micheaux and <u>V</u>. <u>munsoniana</u> Simpson ex Munson, are native to the southeastern U.S.; <u>V</u>. <u>popenoei</u> Fennell occurs only in Mexico.

This paper summarizes North American grape culture, and for muscadine grapes, reviews the native and production ranges, propagation techniques, and relationships between vine growth and insects.

Vitis

subg. Vitis			subg. Muscadinia	
East-Central	West	Tropics	Southeast	Tropics
labrusca	californica	coriacea	rotundifolia	popenoei
aestivalis	arizonica	gigas	munsoniana	
lincecumii	girdiana	bourgaeana		
candicans	acerifolia	cariboea		
rupestris	berlandieri	indica		
riparia	longii			
vulpina	doaniana			
baileyana	monticola			
cinerea	champinii			
palmata	treleasei			
illex				
smalliana				
rufotomentosa				
shuttleworthii				

Figure 1. - North American species of Vitis and their approximate native ranges. (Bailey and Bailey 1976, Comeaux 1984, Brizicky 1965).

NORTH AMERICAN GRAPES

North American grape culture is derived from 4 cultural classes: the American cultivars of V. labrusca L. and V. aestivalis Micheaux grown east of the Rocky Mountains; the V. vinifera cultivars grown in Arizona and California; the French hybrids derived from hybridization of V. vinifera and North American species; and, the muscadine cultivars of V. rotundifolia grown in the southeastern states (Bailey and Bailey 1976).

The fox grape, derived from V. labrusca, and the summer grape, derived from V. aestivalis, are propagated from dormant hardwood cuttings, seed, and bud or scion grafts for fruit production in the eastern U.S. The cold-resistant fox grape, grown from New England to Florida and west to Arkansas, is a strong climber, and its tightly clustered, black to red berries have a strong musky flavor. The mildew resistant summer grape, tolerant to hot climates, is the source of many American cultivars. It is a parent of cultivars, 'Suwanee' and 'Conquistador' (Mortensen 1983) which are grown for fresh market and wine in Florida and Georgia where Pierce's Disease is a limiting factor.

The European grape, derived from V. vinifera, is

widely propagated for fruit production and as an ornamental vine throughout the world. Culture in North America is greatest in California, Arizona, and certain oases in the Northern Mexican Plateau.

The direct producers of French hybrids are derived from interspecific crosses between American species and $V.$ vinifera. The primary diseases of $V.$ vinifera are mildews, and the principal insect pest is the grape phylloxera. Private nurserymen and breeders have conducted most of the breeding of direct producers, and currently half of the vines grown in France are direct producers. These hybrids combine the resistance to pests found in American species with the superior fruit quality of the European grape.

The muscadine grape, $V.$ rotundifolia, is a native North American species which is propagated by ripe soft wood cuttings for fruit production in the southern U.S. from Delaware to Florida and west to Kansas and Mexico (Hedrick 1908). The U.S. census of agriculture (U.S.D.C. 1984) lists grape vine counts for the southeastern states but does not separate bunch from muscadine vines (Fig. 2).

In this region muscadine vines are easier to propagate than $V.$ vinifera because muscadines are not as susceptible to certain insect pests and plant pathogens. The ease of propagation has made the muscadine grape an economically important fruit crop. In addition, it has been crossed with $V.$ vinifera to form insect and disease resistant hybrids (Olmo 1971).

Wild muscadine grapes were harvested by early Americans for fresh fruit, and early cultivars were selected from wild plants (Reimer 1909). In 1811, on Roanoke Island in North Carolina, a sport of the wild muscadine was selected for production of wine grapes. This variety was named Scuppernong after a river near the discovery site. By 1840, North Carolina led the nation in wine production. Production in Georgia surpassed North Carolina in 1850. 'Virginia Dare' a pure muscadine wine won prizes in international competition. The native Scuppernong vines did not have the problems of uneven ripening of fruit, low soluble solids, cold injury, or short vine life which were associated with $V.$ vinifera plantings in the southeastern U.S.

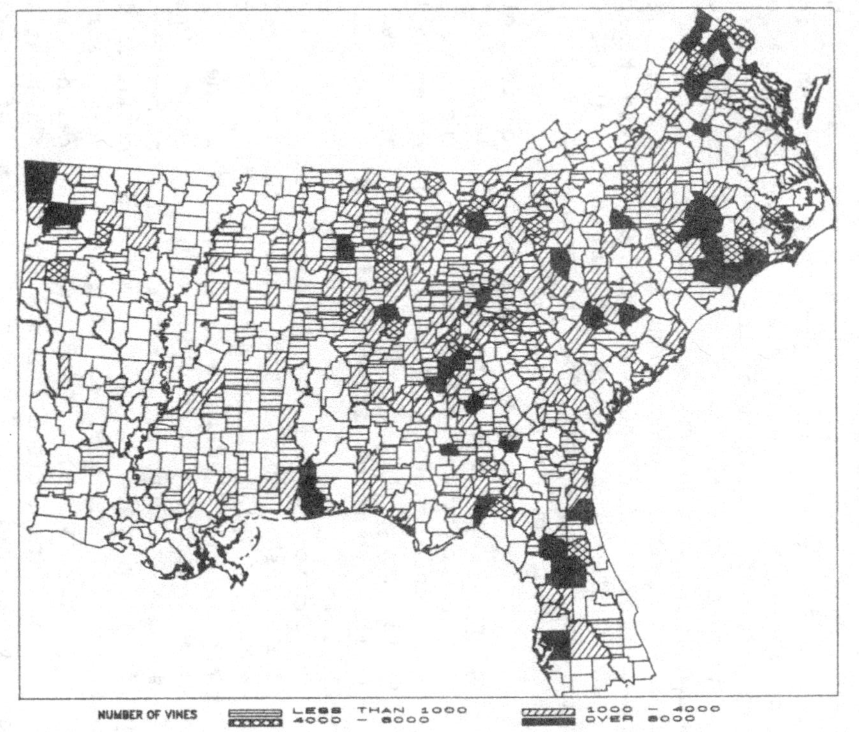

Figure 2. - Muscadine and bunch fresh market and wine grape production is distributed across the southeastern U.S. (U.S. Dept. of Commerce 1984).

Strains of Scuppernong have different fruit characteristics and fruit development times (Woodroof 1934). Controlled hybridization research has developed cultivars with greater berry size, higher berry number per bunch, higher soluble solids, drier stem scars on picked fruit, greater yield and vigor (Lane 1977), and self-fertile flowers (Loomis and Williams 1957). The older cultivars 'Hunt', 'Thomas', and 'Scuppernong' are being replaced by new bronze cultivars, such as 'Carlos', 'Dearing', 'Fry', 'Higgins', 'Magnolia', and 'Welder', and newer black cultivars, including 'Albemarle', 'Chief', 'Cowart', 'Creek', 'Jumbo', 'Magoon', 'Noble', 'Pride', 'Southland', and 'Tarheel'. Self-fertility reduces vineyard layout constraints; black grapes are not preferred when sold in fresh market with bronze grapes; and, a dry stem scar increases berry shelf life and facilitates harvesting of the berries. Selection of cultivars for wine production is dependent on evenness of ripening, flavor, sugar content, and overall performance at a particular vineyard location.

Breeding programs are underway in Florida, Georgia, Mississippi, and North Carolina for new cultivars. The development time for a new cultivar from first cross to release ranges from 12 to 27 years for recent muscadine cultivars. Release is dependent on the value of the berries in fresh fruit or wine markets. The main breeding parents are 'Cowart', 'Fry', 'Magnolia', 'Hunt', 'Higgins', and 'Topsail'. Promising new muscadine cultivars are being tested across the southeastern U.S. in a regional research project. 'Dixie' (Nesbitt et al. 1976), 'Welder' (Mortensen 1977), 'Doreen' (Nesbitt et al. 1982b), 'Triumph', and 'Summit' (Lane 1977, 1980) are bronze, self-fertile cultivars. 'Nesbitt' (Goldy and Nesbitt 1985) and 'Regale' (Nesbitt et al. 1982a) are black, self-fertile cultivars which are also winter hardy.

INSECTS OF MUSCADINE GRAPES

More than 185 insects occurring in the southeastern U.S. feed on grapes. During 1978 and 1979, in a survey of North Carolina grape insects, 109 grape-feeding species were collected. Of these, more than 70% feed on shoots, leaves, or buds; about 20% are fruit or flower feeders; and less than 5% attack the roots. About a third of the species were collected only on muscadine grapes, and just over a fourth were found only on bunch grapes (McGiffen and Neunzig 1985). Three species, the grape root borer, the grape curculio, and the green June beetle, seriously damage vines, fruit, or both each year in some areas where muscadine grapes are commercially grown. Because other grape-feeding insects generally occur in low numbers or cause only minor injury to leaves, they usually do not limit vine growth or fruit production. In the remainder of this section, 24 of the more common insect pests of muscadine grapes are discussed, and less common but related grape-feeding species are mentioned.

As a larva, the grape root borer, _Vitacea polistiformis_ (Harris), a pernicious pest of muscadine grapes, infests the lateral roots and stem base. The eggs are red to brown and are found on the soil surface and on grape leaves during the summer and fall. The cylindrical larva is white to cream-colored with a brown, retractable head. It has short thoracic legs and 2 bands of crochets on the abdominal prolegs. When full grown, the larva is 3-4 cm long. The adult, a brown moth with transparent hind wings and yellow bands on the abdomen, has a wasplike appearance during flight.

Larvae usually feed inside the larger lateral roots or stem base just below the suberized tissue in gouge-shaped wounds. Larvae also tunnel through the soil around the root system (Bambara 1977). Old feeding damage is characterized by irregular or flat surfaces on the roots and pulverized frass packed into the wounds. The damage, caused by direct larval feeding, ranges from a few feeding sites to complete destruction of the root system. When populations are not controlled, vine decline occurs over 4 to 5 seasons with yield losses of 25-50% per year. An infestation typically precedes vine decline by 2 to 3 years.

Adult clearwing moths emerge from the soil from late June to September. Females deposit eggs on the soil surface and on grape leaves within 8 days of adult emergence. Each female lays an average of 354 eggs. First instar larvae enter the soil searching for grape roots. At this stage, predators and dessication contribute to very low survival. However, once the larvae are established in the root, survivorship to adult is very high. Larvae have 6 instars and feed in the roots for 2 or 3 seasons before pupating in silk cocoons near the soil surface. The cocoons are formed in June and are oriented vertically in the soil, often attached to the infested root or stem base. They can be found by scraping away the soil surface around the base of the stem to a distance of 35 cm. Pupae are white to brown depending on development and are 1.5-2.5 cm long (Dutcher and All 1979).

The grape root borer can be controlled by mounding soil over the infested area. This prevents larvae from entering the root zone and pupae from emerging through the mound. The soil surface can also be treated with insecticide to kill larvae as they enter the soil. Adults emerge near harvest and insecticides that can be applied at this time have a low toxicity to adults. Hoeing the soil surface in June and July for weed control above the infested area of each vine will also kill or expose pupae.

The grape leaffolder, _Desmia funeralis_ (Hübner), is a pyralid moth that, as a larva, infests grape foliage, rolling an entire leaf around its body and feeding within the roll. The iridescent eggs are laid on the grape leaves along the veins. Each female lays an average of 199 eggs. The yellow to green cylindrical larvae are gregarious and make webs between leaves during early instars. Later stages roll leaves individually, each larva rolling 1 or 2 leaves and growing through 5 instars. Pupation can occur in the rolled leaf, in leaves webbed together, or in debris on the ground. The emerging adults have black wings with a span of 2-3 cm and 2 pair of white spots on the forewing and 1 white band on each hindwing. The insect has 2 or 3 generations per year on muscadine grapes in the southeastern U.S. Damage incidence is usually low because several parasitic flies and wasps reduce survival of the larvae and pupae. Damage potential and fecundity are high.

Adults and larvae of the grape flea beetle, _Altica chalybea_ Illiger, are foliage feeders on muscadine grapes. Adults are shiny, dark blue beetles, about 4-5 mm long. Larvae, when full grown, are yellow, 7-8 mm long, caterpillarlike insects with black spots and head.

Grape flea beetles pass through one generation per year. Overwintering adults move to the grapevines in April, and females deposit eggs near emerging grape buds. Larvae, most abundant between April and July, feed on grape leaves making ragged feeding holes in the foliage. A new generation of adults is produced from mid June to mid July, and they feed for a short time before overwintering. Damage incidence is spotty within a given vineyard, with the highest populations being found near surrounding wooded sites. The lesser grapevine flea beetle, _Altica woodsi_ Isely, also feeds on muscadine grapes, but it is much less abundant. The adults are smaller than those of _A. chalybea_, and they emerge from hibernation about 3 weeks later.

Two serpentine leafminers, _Phyllocnistis vitegenella_

Clemens and P. vitifoliella Chambers, feed, as larvae, between the upper and lower surfaces of muscadine grape leaves. Pupation sites are at the leaf margins. The adult gracillariid moth, with a wingspan of 0.5 cm, has silvery-white wings with a black spot on each forewing. Feeding only causes significant damage when the mines are formed in emerging foliage, thereby reducing growth. The incidence of high leafminer populations is very low in commercial muscadine vineyards in Georgia and North Carolina because several parasitic wasps reduce survival of the larvae. Populations at low levels, however, are very common. P. vitegenella is the more common of the 2 leafminers in North Carolina.

The grapevine aphid, Aphis illinoisensis Shimer, is very common on new foliage, tendrils and fruit of muscadine grapes. Its life cycle is complex and involves many forms and 2 host plants. In the fall, mated females lay eggs on the stems of Viburnum. Stem mothers hatch from the eggs in the spring and feed on Viburnum. A second parthenogenetic generation, occurring on Viburnum, produces alate viviparous females which migrate to grape. Several generations of apterous and alate summer forms reproduce on the grapevines. For apterous females, the average generation time is 10 days with each female producing an average of 168 nymphs at a rate of 6-10 nymphs per day. Alate females, produced in the fall, migrate to Viburnum where they deposit oviparous females. These mate with alate migrant males, and egg laying follows.

Most adults on grapes are alate or apterous, nearly black, live-bearing females, about 1.5-2.0 mm long. The nymphs, which are found clustered around the adults on the distal ends of new foliage, are green to black. Stunting of terminal stem growth and fruit drop have been attributed to feeding damage by these aphids. In Georgia and North Carolina, low population levels are common, but high population levels are rare. Ants tend the aphids and collect honeydew, preventing the residue from accumulating on the foliage. Parasitic wasps contribute to high mortality during the fall months in Georgia and North Carolina.

The green June beetle, Cotinis nitida (L.), is a large (2 cm long), robust scarab with velvety-green wing covers margined with bronze. Adults feed heavily on ripened fruit, including the berries of muscadine grapes. The grubs, which feed on the roots of grasses, legumes, and other plants grown in organic soils, overwinter in the soil. Adults begin emerging in June and are most abundant in July and August. One generation develops each year.

Two other scarab beetles are pests on muscadine grapes. The Japanese beetle, _Popillia japonica_ Newman, is a metallic green beetle with copper-colored wing covers, about 13 mm long. Adults feed gregariously, skeletonizing the foliage of many plants including muscadine grapes. The life history of the Japanese beetle is similar to that of the green June beetle. Adult emergence occurs from mid June through July. The spotted pelidnota, _Pelidnota punctata_ (L.), is a large (2-3 cm long), robust, tan to rust-colored scarab with 3 black spots on each wing cover. The nocturnal adults occurring from June through September feed on grape foliage causing a ragged appearance.

The grape curculio, _Craponius inaequalis_ (Say), is an important weevil pest of muscadine grapes. The adults are black weevils which measure 3.0 mm long by 2.5 mm wide. They feed in a zigzag pattern on upper leaf surfaces or on fruit petioles. The legless, yellow to white larvae cause considerable damage by feeding on the berry pulp and seeds. The life cycle from egg to adult requires an average of 35 days. Overwintering adults leave hibernation in late May and feed for about 25 days. In mid May, each female deposits single eggs in 1 to 14 berries per day, for an average of 58 days. Larvae feed for about 12 days, and pupate in the ground litter for an average of 19 days. Adults emerge in late summer and feed until their hibernation at the onset of cold weather. The damage incidence and potential are high for this insect, and chemical insecticides applied to the foliage are often required. The adult feeding scars on the upper leaf surfaces are a good indicator of beetle presence. A thorough cleanup of the leaf litter and rotten berries beneath the trellis should reduce weevil density.

The grape berry moth, _Endopiza viteana_ Clemens, a tortricid moth, is an important pest of muscadine grapes that, as a larva, injures the flowers and fruit. Larvae are gray to purple with brown legs and a brown to yellow head. Grape curculio larvae which also inhabit the grape berries can be recognized by their legless condition, white body, and brown head. The grape berry moth has 2 or 3 generations per year in the southeastern U.S. Eggs are laid on the flowers and fruit, and emerging larvae feed and produce webbing near the oviposition site. Each larva feeds on the pulp and seeds of up to 4 berries before it pupates in the foliage or on the ground. Bunch grapes are attacked early in the season, and muscadines are attacked in mid to late summer. Damaged berries have larval entrance holes near the pedicel or between adjacent fruit surfaces, with light webbing and frass around the infested area.

The stink bugs, <u>Nezara</u> <u>viridula</u> L., <u>Acrosternum</u> <u>hilare</u> (Say), and <u>Euschistus</u> <u>servus</u> (Say), are important fruit feeding hemipterans of muscadine grapes. Stink bugs, which have piercing-sucking mouthparts, feed as adults and nymphs on fruit and seeds causing fruit drop and fruit pitting. Adults are generally shield-shaped with the forewings leathery in front and membranous distally. <u>N</u>. <u>viridula</u>, the southern green stink bug, which is yellow-green as an adult, has a broad host range and 4 generations per year. It migrates from one host to the next throughout the season and overwinters as an adult in any hidden place, such as beneath tree bark or under leaf litter. The robust, bright adults of <u>A</u>. <u>hilare</u>, the green stink bug, feed on grape berries in early summer, whereas, adults of <u>E</u>. <u>servus</u>, the brown stink bug, typically feed on berries during the spring and summer. Grape is a feeding host for these stink bugs, but they immigrate from and emigrate to other hosts during the season.

The banded grape bug, <u>Taedia</u> <u>scupeus</u> (Say), and the tarnished plant bug, <u>Lygus</u> <u>lineolaris</u> (Palisot de Beauvois) are grape feeding mirids or plant bugs. Like stink bugs, they have piercing-sucking mouthparts, but they are much smaller (5-6 mm long compared to 15-18 mm long). By their feeding, these bugs cause considerable damage to the buds in the spring. The banded grape bug is an orange and black hemipteran that feeds on blossoms and young fruit causing injury to emerging leaves and blossom clusters. The tarnished plant bug feeds on many fruit, vegetable, and legume species. This bug is found on grapes during early summer and feeds on the buds, blossoms, and new leaves.

A common group of homopterans on muscadine grapes are the grape leafhoppers. These insects breed on grapes but also feed, during the early spring before grape buds break, on various weeds commonly found in the vineyard, such as plantain, dock, wild strawberry, rhubarb, clover, and hollyhock. The more common species, <u>Erythroneura</u> <u>vulnerata</u> Fitch, has 2 to 3 generations per year. Adults overwinter on the vineyard floor and feed on various weeds. They lay the eggs in the leaves and leaf petioles in late spring and the adults and nymphs feed throughout the remainder of the season on grape leaves. Damage is caused when the chlorophyll is removed from the leaf by feeding. The damage appears as a cluster of white spots on the leaf. Further infestations cause leaf death and drop. The leafhoppers can also cause extensive honeydew buildup on the foliage and fruit. Chemical controls are required when 15 to 20 nymphs per leaf are present or before 20% leaf loss occurs. Other common grape leafhoppers occurring in North Carolina are: E. calycula McAtee, E. diva McAtee, E. vitis var. corona McAtee, Zygina illinoiensis (Gillette), and Hymetta balteata McAtee.

Colaspis recurva Blake is the most common of 3 Colaspis beetles known to feed on muscadine grapes. The other 2 are C. floridana Schaeffer and C. carolinensis Blake. These leaf beetles are yellowish-brown with pale stripes, oval in shape, and about 4-5 mm long. They feed on muscadine grape leaves, producing small, somewhat oval holes.

Adults of the southern grape rootworm, Fidia longipes (Melsheimer), are gray beetles, about 4-6 mm long and 2-3 mm wide, with long legs and threadlike antennae. These beetles feed on muscadine grape leaves, producing chainlike series of small holes. They are most abundant during May and June. The closely-related grape rootworm (Fidia viticida Walsh) also feeds on muscadine grapes, but it is less abundant. Larval stages of the grape rootworm are known to weaken or kill bunch grape vines by feeding on roots. Such damage has not been documented for either rootworm species on muscadine grapes.

Heliozela sp. near aesella Chambers is a gall-forming moth. In the spring, irregular swellings are produced along the sides of young shoots, leaf petioles, and flower stems of muscadine grapes. The tiny Heliozela caterpillars feed within the galls, producing channels which weaken plant structures and contribute to breakage. Mature larvae drop from the vine in tubes constructed about themselves and pupate in soil or debris. No leaf galls of H. aesella Chambers, a close relative, were found on muscadine grapes.

The greater grapevine looper is a slender pale green or brown caterpillar, approximately 4 to 5 cm long when full grown, which resembles a tendril, petiole, or twig. These loopers feed on the leaves of muscadine grapes, sometimes consuming entire leaf blades. Although there are 2 or more generations per year, most injury occurs during the spring months and is generally minor. The adult is an attractive yellow moth, about 3.5 cm long. The grapevine looper or lesser grapevine looper, Eulithis diversilineata Hübner, also fits the above description but is less common in the southeastern U.S. on muscadine grapes.

Muscadine grapes are attacked by 3 skeletonizing caterpillars. The most common of these is Acoloithus falsarius Clemens. The other 2 are A. novaricus Barnes &

84

McDunnough and the grapeleaf skeletonizer, <u>Harrisina</u> <u>americana</u> Guérin Méneville. Adults of all 3 insects are small, smoky-gray to black, day-flying moths which have orange or red collars. Full-grown larvae (about 10-15 mm) are hairy, yellow or pale brown caterpillars, marked with brown stripes, bands or spots. Young skeletonizers feed gregariously in orderly rows on leaf surfaces; later stages disperse. The caterpillars are most abundant in June and July. A second, smaller generation occurs in late summer. These skeletonizers overwinter as pupae in silk cocoons.

A common sphinx moth with a grape-feeding larva is the Virginiacreeper sphinx, <u>Darapsa</u> <u>myron</u> (Cramer), also known as the grapevine sphinx or hog caterpillar. The larval hornworm grows to a full length of approximately 4 cm and ranges in color from yellowish-brown to green or bluish-green. Two generations occur in North Carolina, and caterpillars can be found feeding on the leaves of muscadine grapes from spring through late summer. Full-grown larvae are only occasionally encountered. Apparently, predation and parasitism effectively control populations.

Two more leaf-feeding caterpillars occurring on muscadine grapes are the eightspotted forester, <u>Alypia</u> <u>octomaculata</u> (Fabricius), and the copper underwing, <u>Amphipyra</u> <u>pyramidoides</u> Guenée. The common names refer to the adult stages of these insects. The former is a black moth, marked with 8 white spots, and the latter is a brown moth with copper-colored underwings. Eightspotted forester larvae are brownish-yellow when small, but later stages (1-3 cm) are striped with black, white and orange. With at least 2 generations occurring each year, the forester caterpillars can be found from spring through summer. Copper underwing larvae, also known as pyramidal fruitworms, are apple green in color, about 3-4 cm when full-grown, and distinguished by a posterior hump. As young larvae, these caterpillars feed in early spring on tiny new leaves. The pyramidal fruitworm has 1 generation per year.

INSECT-PLANT RELATIONS

Native <u>V</u>. <u>rotundifolia</u> vines occur along roadsides, in wooded areas, and adjacent to many commercial vineyards in the southeastern U.S. These vines harbor all the indigenous insects listed in Table 1, as well as, their parasites and predators. High populations of grape-feeding insects are

Table 1. General population trends** [generation time as no. gen./yr, fecundity as no. eggs (nymphs) per female, damage incidence (D.I.), and damage potential (D.P.)] and general type of control [biological (bio.), chemical insecticide (chem.) or none] for insects commonly found feeding on muscadine grape in the southeastern United States.

Species - Common Name	Gen./yr.	Fecundity	D.I.	D.P.	Control
Fruit feeders					
Cotinis nitida (L.) -					
green June beetle [Scarabaeidae]	1/2	*	high+	high	chem.
Craponius inaequalis (Say) -					
grape curculio [Curculionidae]	1	590	high+	high	chem.
Endopiza viteana Clemens -					
grape berry moth [Tortricidae]	2 or 3	212	mod.+	high	bio./chem.
Taedia scrupeus (Say) -					
banded grape bug [Miridae]	?	?	low	mod.	none
Acrosternum hilare (Say) -					
green stink bug [Pentatomidae]	?	*	mod.	mod.	none
Euschistus servus (Say) -					
brown stink bug [Pentatomidae]	?	*	mod.	mod.	none
Nezara viridula L.					
southern green stink bug[1] [Pentatomidae]	4	*	high	mod.	none
Leaf and shoot feeders					
Popillia japonica Newman -					
Japanese beetle[2] [Scarabaeidae]	1	*	mod.	high	chem.
Erythroneura vulnerata Fitch -					
grape leafhopper [Cicadellidae]	2 or 3	75-100	low-mod.	high	chem.
Desmia funeralis (Hübner) -					
grape leaffolder [Pyralidae]	2 or 3	199	low	high	bio.
Acoloithus falsarius Clemens -					
a skeletonizer [Zygaenidae]	2	?	mod.	high	none

Insect					
Heliozela sp. near aesella Chambers –					
a gall former [Heliozelidae]	1	?	low	high	none
Alypia octomaculata (Fabricus) –					
eightspotted forester [Noctuidae]	2 or 3	?	low	high	none
Aphis illinoisensis Shimer –					
grapevine aphid [Aphididae]	many	168	high+	mod.	bio./chem.
Altica chalybea Illiger –					
grape flea beetle [Chrysomelidae]	1	20-100	low	mod.	none
Lygus lineolaris					
(Palisot de Beauvois) –					
tarnished plant bug [Cicadellidae]	3 to 5	*	low	low	none
Pelidnota punctata (L.) –					
spotted pelidnota [Scarabaeidae]	1/2	*	high+	low	none
Phyllocnistis vitegenella					
Clemens – a leafminer [Gracillariidae]	1	?	mod.	low	bio.
Darapsa myron (Cramer) –					
Virginiacreeper sphinx [Sphingidae]	2	?	low	low	chem.
Colaspis recurva Blake –					
a leaf beetle [Chrysomelidae]	1	*	low	low	none
Eulithis gracilineata Guenee –					
greater gravevine looper [Geometridae]	2 or 3	?	low	low	none
Amphipyra pyramidoides (Guenee –					
pyramidal fruitworm [Noctuidae]	1	?	low	low	none
Root feeders					
Vitacea polistiformis (Harris) –					
grape root borer [Sesiidae]	1/2 or 1/3	354	high	high	chem.
Fidia longipes (Melsheimer)–					
southern green rootworm [Chrysomelidae]	1	?	mod.	mod.	none

* Muscadine grape is a feeding and not a breeding host for these insects. ? Value unknown.
+ D.I. lower in North Carolina. 1 Probably an introduced species. 2 Introduced species.

rare on muscadine grapes. In general, foliage remains intact and green throughout the summer, and most leaves naturally senesce in the autumn. From 1978 through 1983 in North Carolina, heavy and widespread leaf feeding with subsequent defoliation of wild muscadine vines was observed only once. This damage was caused by *Acoloithus falsarius* and *A. novaricus*, two species related to the grapeleaf skeletonizer, *Harrisina americana* (Guérin Méneville) and the western grapeleaf skeletonizer, *H. brillians* Barnes & McDunnough, serious pests of native bunch grapes and western *V. vinifera* grapes, respectively.

Grape phylloxera and Pierce's disease are lethal to many commercial cultivars of *V. vinifera* and *V. labrusca* (Olmo 1971). If not treated, native American bunch grapes are often severely damaged by grape berry moth larvae and Japanese beetle adults. Cecidomyiid galls in numerous forms are common on bunch grapes, sometimes causing substantial plant deformity or bud drop. McGiffen and Neunzig (1985) reared seven species from North Carolina bunch grapes in 1978. In contrast, muscadine grapes are tolerant to grape phylloxera, and Pierce's disease limits production but is not lethal (Mortensen 1986). Attacks by grape berry moths and Japanese beetles do not necessarily require control measures. In North Carolina, during 1978 and 1979, three untreated muscadine vineyards were monitored by McGiffen and Neunzig (1985). In spite of the presence of grape berry moths, Japanese beetles, and an assortment of other grape pests, fruit yield and quality at harvest were good. None of these vineyards showed any evidence of grape root borer presence. Finally, muscadine grapes seem to be resistant to most forms of leaf galls, including those of numerous cecidomyiids, grape phylloxera, and *Heliozela aesella* Chambers. Only one type of cecidomyiid leaf gall, a small round swelling, was observed by McGiffen and Neunzig (1985) on muscadine grapes.

Recent evidence (Gopakumar et al. 1977) indicates that grapevines produce high levels of compounds which mimic the activity of juvenile hormone on insects which feed on the vine. The determination of the juvenomimetic activity of extracts from muscadine grape cultivars might indicate a cause of the apparent tolerance of these plants to insect attack. Gopakumar et al. found activity levels 2X to 18X higher in grape than in other fruit, or tree crops.

The commercial producer of muscadine grapes in Georgia usually applies 3 fungicide and 2 insecticide sprays each season. The fungicide sprays protect the berries from ripe rot (*Gleomerella cingulata*), bitter rot (*Melanconium*

88

fugligineum), and macrophoma rot (Botryosphaeria dothidea)
and the insecticide sprays are applied to the soil for grape
root borer control (All et al. 1985). Other insect pests
are monitored and treated as needed (Table 1). Grape
leaffolder is easily controlled by foliar sprays of the
Bacillus thuringiensis (Biever and Hostetter 1975).
Japanese beetle is usually controlled with a carbaryl spray
in areas where the beetle occurs. Grape leafhoppers cause
considerable damage to V. vinifera (Jensen et al. 1973) and
V. labrusca (Jubb et al. 1978) grapes throughout the U.S.
but occur on V. rotundifolia at levels that seldom require
chemical control. Insecticide-induced insect problems,
such as resurgence of minor pests or tolerance to
insecticides have not occurred in muscadine grape culture.

Current muscadine grape breeding programs do not have
insect resistance as a major goal. Screening cultivars for
insect resistance would not be difficult given the ease of
clonal propagation by rooting softwood cuttings (Goode and
Lane 1983), the relatively short time from rooting to vine
establishment of 3 years, and the diversity of parent plant
material in woodlots. Pierce's disease was the major
limiting factor in muscadine grape production in Florida
where Mortensen developed resistant commercial cultivars,
such as 'Southland' and 'Dixie' by crossing cultivars with
native vines (Nesbitt et al. 1976) and by developing
'Welder' from a dooryard vine found to be resistant and to
having good fruit characteristics.

The relative tolerances to insect and disease pests
found in V. rotundifolia and V. munsoniana are valuable
resources for the southeastern U.S. Crosses of V.
rotundifolia with V. munsoniana may produce insect
resistant, commercially acceptable cultivars.
Marketability and quality consistency of muscadine grape
products have been improved through development of
techniques to produce clearer juices and wines (G. M.
Brooks, personal communication). The goals for entomology
in muscadine grape production are: the retention of
natural resistance to insect attack through augmentation of
breeding programs with host plant resistance to insects,
and the development of chemical controls for pernicious
insect pests that do not disrupt the biological control
agents of minor pests.

LITERATURE CITED

All, J. N., J. D. Dutcher, M. C. Saunders, and U. E. Brady. 1985. Prevention strategies for grape root borer (Lepidoptera: Sesiidae) infestations in concord grape vineyards. J. Econ. Entomol. 78:666-670.

Bailey, L. H. and Bailey, E. Z. 1976. Hortus Third. MacMillan Publ. Co., Inc. N. Y. pp. 1162-1163.

Bambara, Stephen B. 1977. Biology of the grape root borer, _Viticea polistiformis_, in eastern North Carolina with descriptions of immature stages [Lepidoptera: Sesiidae]. M.S. Thesis. North Carolina State Univ. Raleigh.

Biever, K. D. and D. L. Hostettler. 1975. _Bacillus thuringiensis_: Against Lepidopterous pests of wine grapes in Missouri. J. Econ. Entomol. 68:66-68.

Brizicky, G. K. 1965. The genera of Vitaceae in the southeastern United States. J. Arnold Arboretum. 46:48-67.

Comeaux, B. L. 1984. Taxonomic studies of certain native grapes of the eastern United States. M. S. Thesis. North Carolina State Univ. Raleigh.

Dutcher, J. D. and J. N. All. 1979. Biology and Control of the grape root borer in Concord grape vineyards. Georgia Agric. Exp. Station. Research Bull. 232. 18 p.

Goldy, R. G. and W. B. Nesbitt. 1985. 'Nesbitt' muscadine grape. HortScience 20:777.

Goode, Jr., D. Z., and R. P. Lane. 1983. Rooting leafy muscadine grape cuttings. HortScience 18:944-946.

Gopakumar, B., B. Ambika, and V. K. K. Prabhu. 1977. Juvenomimetic activity in some south Indian plants and the probable cause of this activity in _Morus alba_. Entomon 2:259-261.

Hedrick, U. P. 1908. The grapes of New York. N. Y. State Agric. Exp. Sta., Geneva, Rpt. No. 1907.

Jensen, F., Don Flaherty, Curtis D. Lynn, R. L. Doutt. 1973. Grape leafhopper. Univ. Calif. Agric. Extension Service Bulletin-Grape Pest Management in the San Joaquin Valley.

Jubb, G. L., T. H. Obourn, and D. H. Peterson. 1978. Pilot pest management program for grapes in Erie County, Pnnsylvania. J. Econ. Entomol. 71:913-916.

Lane, R. P. 1977. 'Summit' muscadine grape. HortScience 12:588.

Lane, R. P. 1980. 'Triumph' muscadine grape. HortScience 15:322.

Loomis, N. H., and C. F. Williams. 1957. A new flower type of the muscadine grape. J. Hered. 48: 294-304.

McGiffen, K. C., and H. H. Neunzig. 1985. A guide to the identification and biology of insects feeding on muscadine and bunch grapes in North Carolina. N. C. Agric. Res. Service Bull. 470, 93pp.

Mortensen, J. A. 1977. 'Welder' muscadine grape. HortScience 12:267-268.

Mortensen, J. A. 1983. 'Suwanee' and 'Conquistador' grapes. HortScience 18:767-769.

Mortensen, J. A. 1986. Grape varieties, rootstocks, and propagation. Proc. Viticultural Science Symp. Florida A&M Univ. pp. 13-25.

Nesbitt, W. B., D. E. Carroll, Jr., J. P. Overcash, and B. J. Stojanovic. 1982a. Regale' muscadine grape. HortScience 17:276-278.

Nesbitt, W. B., D. E. Carroll, Jr., J. P. Overcash, C. P. Hedgewood, Jr., and B. J. Stojanovic. 1982b. 'Doreen' muscadine grape. HortScience 17:278.

Nesbitt, W. B., V. H. Underwood, and J. A. Mortensen. 1976. 'Dixie' grape. HortScience 11:520-521.

Olmo, H. P. 1971. Vinifera x rotundifolia hybrids as wine grapes. Amer. J. Enol. Viticul. 22:87-91.

Pratt, C. 1971. Reproductive anatomy of cultivated grapes - a review. Amer. J. Enol. Viticul. 22:92-109.

Pratt, C. 1974. Vegetative anatomy of cultivated grapes - a review. Amer. J. Enol. Viticul. 25:131-150.

Reimer, F. C. 1909. Scuppernong and other muscadine grapes: origin and importance. N. C. Agric. Exp. Sta. Bull. No. 201.

Woodroof, J. G. 1934. Five strains of the Scuppernong variety of the muscadine grapes. Proc. Amer. Soc. Hort. Sci. 32:384.385.

U. S. Department of Commerce. 1984. 1982 Census of Agricultural Vol. 1 Geographic Area Series. Fruits and Nuts: Grapes. U. S. Gov't Printing Office. Washington D.C.

4

Entomology of the Strawberry

C. H. Shanks, Jr., and T. M. Sjulin

The cultivated strawberry, <u>Fragaria</u> x <u>ananassa</u> Duchn., is a hybrid of three American species: <u>F. virginiana</u> Duchn., <u>F. virginiana</u> (L.) Duchn. ssp. <u>glauca</u> Staudt, and <u>F. chiloensis</u> (L.) Duchn. <u>F. virginiana</u> is native to the meadows of eastern North America from Hudson Bay to Georgia to Louisiana to the Dakotas. <u>F. virginiana glauca</u>, which occurs as a parent in a few <u>F.</u> x <u>ananassa</u> cultivars, grows in an area bounded by northern New Mexico, California beaches, Alaska, and Montana (Darrow 1966). <u>F. chiloensis</u> occurs on the beaches and Andes Mountains of Chile, high on mountains of Hawaii, and on the Pacific Ocean beaches from near Santa Barbara, California north to the Aleutian Peninsula of Alaska.

All the above species are octoploid (2n = 56). They will all freely intercross and produce viable seeds (Scott et al. 1972). All three species are quite variable (Hildreth and Powers 1941; Darrow 1953, 1966; Scott 1959, Bringhurst et al. 1977; Hancock and Bringhurst 1979). These traits permit the utilization of a vast gene pool for developing varieties with many different desirable characteristics such as fruit quality, winter hardiness, and pest resistance.

There is very little recorded in the literature on the insect and mite pests of wild octoploid strawberries. While collecting hundreds of <u>F. chiloensis</u> plants from California, Oregon, and Washington beaches, we have never observed any of the insects or mites that are pests of cultivated strawberries. Schaefers and Allen (1962) reported collecting <u>C. fragaefolii</u> from <u>F. chiloensis</u> on the beaches. Miller (1951, 1953) found viruses in wild <u>F. chiloensis</u> and <u>F. virginiana glauca</u> in Oregon and California. Also, Marcus (1952) found viruses in wild

91

Table 1. Host range of several major arthropod pests of strawberry[a].

Common name	Scientific name	Host range
Cyclamen mite	*Steneotarsonemus pallidus* Banks	Wide
Spider mites	*Tetranychus* spp.	Wide
Lygus bugs	*Lygus* spp.	Wide
Strawberry pameras	*Pachybrachius* spp.	Wide
Meadow spittlebug	*Philaenus spumarius* (L.)	Wide
Potato leafhopper	*Empoasca fabae* (Harris)	Wide
Strawberry aphid	*Chaetosiphon* spp.	Strawberry, *Rosa*, *Potentilla*
Strawberry root aphid	*Aphis forbesi* Weed	Strawberry
Strawberry leafroller	*Ancyclis comptana fragariae* (Walsh & Riley)	Strawberry *Rubus* spp.
Garden tortrix	*Clepsis peritana* (Clemens)	Wide
Strawberry crown miner	*Aristotelia fragaria* Busck	Strawberry
Root weevils	*Otiorhynchus* spp., *Nemocestes* spp.	Wide
Strawberry bud weevil	*Anthonomus signatus* Say	Strawberry, *Rubus*, *Potentilla*
Strawberry crown borer	*Tyloderma fragaria* (Riley)	Strawberry, *Potentilla*
White grubs	*Phyllophaga* spp.	Wide
Strawberry root worms	*Paria* spp.	Wide

[a] List of pests from Schaefers (1981).

strawberries in eastern United States. Names of viruses were not given but it was implied that they were vector-borne, which indicated that aphids occurred on them at least in small numbers.

Few of the insect and mite pests of strawberry are restricted in their host range to strawberry. Of the 16 arthropods designated as major pests of strawberry by Schaefers (1981) (Table 1), only the strawberry root aphid, _Aphis forbesi_ Weed, and strawberry crown miner, _Aristotelia fragaria_ Busck, seem to have no other recorded hosts (Metcalf et al. 1962). The strawberry aphids are found mainly on _Fragaria_ under natural conditions (Dicker 1952). However, they have been recorded on the closely related genus, _Potentilla_, by some observers (Schaefers and Allen 1962), on cultivated rose (Thomas and Jacob 1940) and a wild rose, _Rosa_ sp. (Shanks, unpublished data). All of these plant genera belong to the family Rosaceae. Strawberry crown borer, _Tyloderma fragariae_ (Riley), also is restricted to strawberry and _Potentilla_ (Metcalf et al. 1962). Strawberry leafroller, _Ancyclis comptana fragariae_ (Walsh and Riley), and strawberry bud weevil, _Anthonomus signatus_ Say, are restricted to strawberry and _Rubus_ spp., both rosaceous. Presumably, wild strawberries are important hosts for those insects who are restricted to _Fragaria_, _Potentilla_, and _Rubus_. However, there are few records to verify this.

Wild strawberries in their natural settings do not appear to be heavily infested or damaged by insects or mites. For example, Shanks (unpublished data) could find no strawberry aphids on wild _F. virginiana glauca_ plants growing near a commercial strawberry field. However, when plants were transplanted into the field and grown like commercial cultivars, they became heavily infested by the aphid. When cultivated, strawberries are attacked and damaged by a large number of insects and mites. Schaefers (1981) listed 18 genera and species of major importance and 34 others of minor importance. Many of the latter can become very important regionally or to individual growers. Vegetation around and among the wild plants may harbor predators that serve to keep pests in check. The cool, wet climate, where many of the populations of _F. chiloensis_ exist, is not conducive to development of populations of some of the pests such as spider mites. Also, the soil types where some of the _F. chiloensis_ clones grow would not be suitable for some soil insects, e.g., root weevil larvae.

F. chiloensis is almost the only wild octoploid species recognized as a source of resistance to insect and mite pests. No such sources of resistance have been reported for _F. virginiana_ or _F. virginiana glauca_, except that Kishaba et al. (1972) reported one clone of _F. virginiana glauca_ was highly resistant to the twospotted spider mite. Both of the latter two species have been reported to vary in disease resistance, hardiness, and fruit characters (Hildreth and Powers 1941, Powers 1945,

Darrow and Scott 1947, Scott 1959, Scott et al. 1972). Clones of wild F. chiloensis have been shown to vary greatly morphologically and physiologically (Hancock and Bringhurst 1979). Some clones also have been shown to have resistance to the two most important root diseases of strawberry, red stele and Verticillium wilt (Waldo 1953, Bringhurst et al. 1966). Resistance to red stele disease is readily transmitted to progeny when the 'Del Norte' clone of F. chiloensis is crossed with F. x ananassa (Waldo 1953, Scott and Draper 1976).

Resistance to the strawberry aphid was found in F. chiloensis 'Del Norte' and the resistance was heritable when 'Del Norte' was crossed with F. x ananassa cultivars (Shanks and Barritt 1974). The resistance persisted after the first backcross to F. x ananassa but fruit quality was only fair (Barritt and Shanks 1980). Continued backcrossing to F. x ananassa results in improved fruit quality with a reduced level of aphid resistance (Sjulin and Shanks, unpublished). However, a later breeding experiment [F. chiloensis 'Del Norte' x F. x ananassa 'Linn'] x F. x ananassa 'BC 69-5-34' produced one seedling with very high vigor, resistance to red stele and strawberry aphid, and high quality fruit.

Crock et al. (1982) identified 24 more clones of F. chiloensis as being as resistant to strawberry aphid as 'Del Norte'. They also showed that F. x ananassa cultivars may differ greatly in their susceptibility to the strawberry aphid. This offers the possibility of greatly increasing the gene pool for aphid resistance. Promising F_1 or first backcross seedlings from crosses between the F. chiloensis clones and F. x ananassa clones could be intercrossed to quickly recover both fruit quality and aphid resistance.

Adult black vine weevils, Otiorhynchus sulcatus (F.), and strawberry root weevils, O. ovatus (L.), fed less on leaves of certain clones of F. chiloensis, which resulted in lower fecundity (Shanks et al. 1984). The resistance is due to denser pubescence on the undersides of the leaves (Doss and Shanks 1988, Doss et al. 1987). It is unlikely that this is due to natural selection for resistance to adult weevils because most of the clones come from sites where root weevils probably would not commonly occur. The resistant clones are being crossed with F. x ananassa cultivars in an effort to develop root weevil-resistant strawberry cultivars. A breeding plan similar to that described for strawberry aphid will be followed.

Resistance to spider mites, <u>Tetranychus</u> spp., in the form of lower populations, has been found in <u>F</u>. x <u>ananassa</u> cultivars (Chaplin et al. 1968, Kishaba et al. 1972, Shanks and Barritt 1975), in a clone of <u>F</u>. <u>virginiana glauca</u> (Kishaba et al. 1972), and in <u>F</u>. <u>chiloensis</u> clones (Shanks and Barritt 1984). Tolerance to spider mites was reported by Schuster et al. (1980). These traits are heritable (Chaplin et al. 1968, Shanks and Barritt 1980, Barritt and Shanks 1981). There is so much spider mite-resistant germ plasm available, even in <u>F</u>. x <u>ananassa</u>, it should be very possible to develop strawberry cultivars that are resistant to this very important pest.

As with root weevils, resistance to spider mites in <u>F</u>. <u>chiloensis</u> probably is not due to natural selection. Many of the resistant clones were collected from cool moist environments where spider mites are not likely to thrive.

A clone of <u>F</u>. <u>chiloensis</u> was reported to be relatively resistant to the strawberry bud weevil (Darrow et al. 1933). This resistance was never bred into cultivated strawberries. Again, the resistance to this weevil could not be due to natural selection because the weevil occurs in eastern United States and <u>F</u>. <u>chiloensis</u> is native to the Pacific Coast.

There are no reports in the literature of <u>F</u>. <u>virginiana</u> clones being resistant to any arthropod pests. Since the species has a very wide geographical distribution, it seems likely that there is great genetic diversity that would include arthropod-resistant clones. It is possible that some pest resistance in current cultivars comes from <u>F</u>. <u>virginiana</u> in their ancestry.

In summary, most reported strawberry resistance to arthropod pests is derived from wild <u>F</u>. <u>chiloensis</u> clones. The resistance is not generally a result of natural selection. This observation is similar to that of Harris (1975) who noted that there are many reports of resistance to insect attack in plants that evolved in the absence of the insect to which they are resistant. It is possible that the plant factors which confer resistance to insects actually evolved to benefit the plant in other ways.

Literature Cited

Barritt, B. H. and C. H. Shanks, Jr. 1980. Breeding strawberries for resistance to the aphids <u>Chaetosiphon fragaefolii</u> and <u>C</u>. <u>thomasi</u>. HortScience 15:287-288.

Barritt, B. H. and C. H. Shanks, Jr. 1981. Parent selection in breeding strawberries resistant to twospotted spider mites. HortScience. 16:323-324.

Bringhurst, R. S., J. F. Hancock, and V. Voth. 1977. The beach strawberry, an important natural resource. Calif. Agr. 31(9):10.

Bringhurst, R. S., S. Wilhelm, and V. Voth. 1966. Verticillium wilt resistance in natural populations of _Fragaria chiloensis_ in California. Phytopathology 56: 219-222.

Chaplin, C. E., L. P. Stoltz, and J. G. Rodriquez. 1968. The inheritance of resistance to the two-spotted spider mite _Tetranychus urticae_ Koch in strawberries. Proc. Amer. Soc. Hort. Sci. 92:376-380.

Crock, J. E., C. H. Shanks, Jr. and B. H. Barritt. 1982. Resistance in _Fragaria chiloensis_ and _F._ x _ananassa_ to the aphids _Chaetosiphon_ fragaefolii and _C. thomasi_. HortScience 17:959-960.

Darrow, G. M. 1953. Strawberries in Mexico, Central America, Columbia, and Ecuador. Ceiba 3:179-185.

Darrow, G. M. 1966. The strawberry. Holt, Rinehard and Winston, New York.

Darrow, G. M. and D. H. Scott. 1947. Breeding for cold hardiness of strawberry flowers. Proc. Amer. Soc. Hort. Sci. 50:239-242.

Darrow, G. M., G. F. Waldo, and C. E. Schuster. 1933. Twelve years of strawberry breeding. J. Hered. 24:391-402.

Dicker, G. H. L. 1952. The biology of the strawberry aphid, _Pentatrichopus fragaefolii_ (Cock.), with special reference to the winged form. J. Hort. Sci. 27:151-178.

Doss, R. P., C. H. Shanks, Jr. 1988. The influence of leaf pubescence on the resistance of selected clones of beach strawberry (_Fragaria chiloensis_ (L.) Duchesne) to adult black vine weevils _Otiorhynchus sulcatus_ F.) Sci. Hort. 34:47-54.

Doss, R. P., C. H. Shanks, Jr., J. D. Chamberlain, and J. K. L. Garth. 1987. The role of leaf hairs in the resistance of a clone of beach strawberry, _Fragaria chiloensis_, to feeding by adult black vine weevil, _Otiorhynchus sulcatus_ (Coleoptera: Curculionidae). Environ. Entomol. 16:764-768.

Hancock, J. F. Jr., and R. S. Bringhurst. 1979. Ecological differentiation in perennial octoploid species of _Fragaria_. Amer. J. Bot. 66:367-375.

Harris, M. K. 1975. Allopatric resistance: searching for sources of insect resistance for use in agriculture. Environ. Entomol. 4:661-669.

Hildreth, A. C. and L. Powers. 1941. The Rocky Mountain strawberry as a source of hardiness. Proc. Amer. Soc. Hort. Sci. 38:410-412.

Kishaba, A. N., V. Voth, A. F. Howland, R. S. Bringhurst, and H. H. Toba. 1972. Twospotted spider mite resistance in California strawberries. J. Econ. Entomol. 65:117-119.

Marcus, C. P., Jr. 1952. Survey for virus-infected wild strawberry plants in eastern United States. Plant Dis. Rep. 36:353-354.

Metcalf, C. L., W. P. Flint, and R. L. Metcalf. 1962. Destructive and useful insects. McGraw-Hill Book Co., New York.

Miller, P. W. 1951. Wild strawberries as a source of strawberry virus infection. Plant Dis. Rep. 35:129-130.

Miller, P. W. 1953. Tests of wild strawberries from California for virus infection. Ibid 37:20.

Schaefers, G. A. 1981. Pest management systems for strawberry insects. In D. Pimental, ed. CRC Handbook of Pest Management in Agriculture, Vol. III. CRC Press, Boca Raton, FL. Pp. 377-393.

Schaefers, G. A., and W. W. Allen. 1962. Biology of the strawberry aphids, _Pentatrichopus_ _fragaefolii_ (Cockerell) and _P._ _thomasi_ Hille Ris Lambers, in California. Hilgardia 32:393-431.

Schuster, D. J., J. F. Price, F. G. Martin, C. M. Howard, and E. E. Albregts. 1980. Tolerance of strawberry cultivars to twospotted spider mites in Florida. J. Econ. Entomol. 73:52-54.

Scott, D. H. 1959. Size, firmness, and time of ripening of fruit of seedlings of _Fragaria virginiana_ Duchn. crossed with cultivated, strawberry varieties. Proc. Amer. Soc. Hort. Sci. 74:388-393.

Scott, D. H., A. D. Draper, and L. W. Greeley. 1972. Interspecific hybridization in octoploid strawberries. HortScience 7:382-384.

Scott, D. H., and A. D. Draper. 1976. Reaction of strawberry seedlings inoculated with race composites of _Phytophthora fragariae_ Hickman. J. Amer. Soc. Hort. Sci. 101:355-358.

Shanks, C. H., Jr., and B. H. Barritt. 1974. _Fragaria chiloensis_ clones resistant to the strawberry aphid. HortScience. 9:202-203.

Shanks, C. H., Jr., and B. H. Barritt. 1975. Resistance of strawberries to the twospotted spider mite. J. Econ. Entomol. 68:7-10.

Shanks, C. H., Jr. and B. H. Barritt. 1980. Twospotted spider mite resistance of Washington strawberries. J. Econ. Entomol. 73:419-423.

Shanks, C. H., Jr. D. L. Chase and J. D. Chamberlain. 1984. Resistance to clones of wild strawberry, Fragaria chiloensis, to adult Otiorhynchus sulcatus and O. ovatus (Coleoptera: Curculionidae) Environ. Entomol. 13:1042-1045.

Shanks, C. H., Jr. and B. H. Barritt. 1984. Resistance of Fragaria chiloensis clones to the twospotted spider mite. HortScience. 19:640-641.

Thomas, I., and F. H. Jacob. 1940. The strawberry aphid, Pentatrichopus (Capitophorus) fragariae Theob., with notes on P. potentillae Walk. and P. tetrahodus Walk. Ann. Appl. Biol. 27:234-247.

Waldo, G. F. 1953. Sources of red stele root disease resistance in breeding strawberries in Oregon. Plant Disease Reptr. 37:236-242.

5

Entomology of Rabbiteye Blueberries, *Vaccinium ashei* Reade, in Southeastern United States

J. A. Payne, D. J. Horton, and A. Amis

The rabbiteye blueberry, Vaccinium ashei Reade, is native to the acid, infertile, imperfectly drained soils of the southeastern United States. Major wild populations still exist along the Satilla River in southeast Georgia, the Suwanee River in north central Florida, and the Yellow River in south Alabama and west Florida (Fig. 1). The rabbiteye blueberry is the most recent of the major fruit crops to be brought under cultivation, having been domesticated entirely in the twentieth century. The first commercial rabbiteye blueberry planting from native plants was made near Crestview, Florida about 1893 (Lyrene and Sherman 1979). Little research was done with this fruit until 1926, when the first planting of five plants each of twelve native selections from west Florida was made at the Georgia Coastal Plain Experiment Station (Brightwell and Woodard 1955). The first commercial rabbiteye blueberry enterprise using improved cultivars from the joint U.S. Department of Agriculture-University of Georgia breeding program was planted in 1955 near Savannah, Georgia (Austin 1979). In 1986, there were 5085 acres of improved cultivars in production in 10 southeastern and mid-south states (Fig. 2), with Georgia having approximately half of the total acreage (Table 1).

Phipps (1930) catalogued over 292 species of insects and mites on blueberries and huckleberries; however, rabbiteye blueberries were not included in this survey. Marucci (1966) compiled ratings of the 24 most important blueberry insects in the various growing areas, but the rabbiteye blueberry growing area was not surveyed. Galletta (1975), in his chapter on breeding blueberries and cranberries, lists no potentially serious insects of

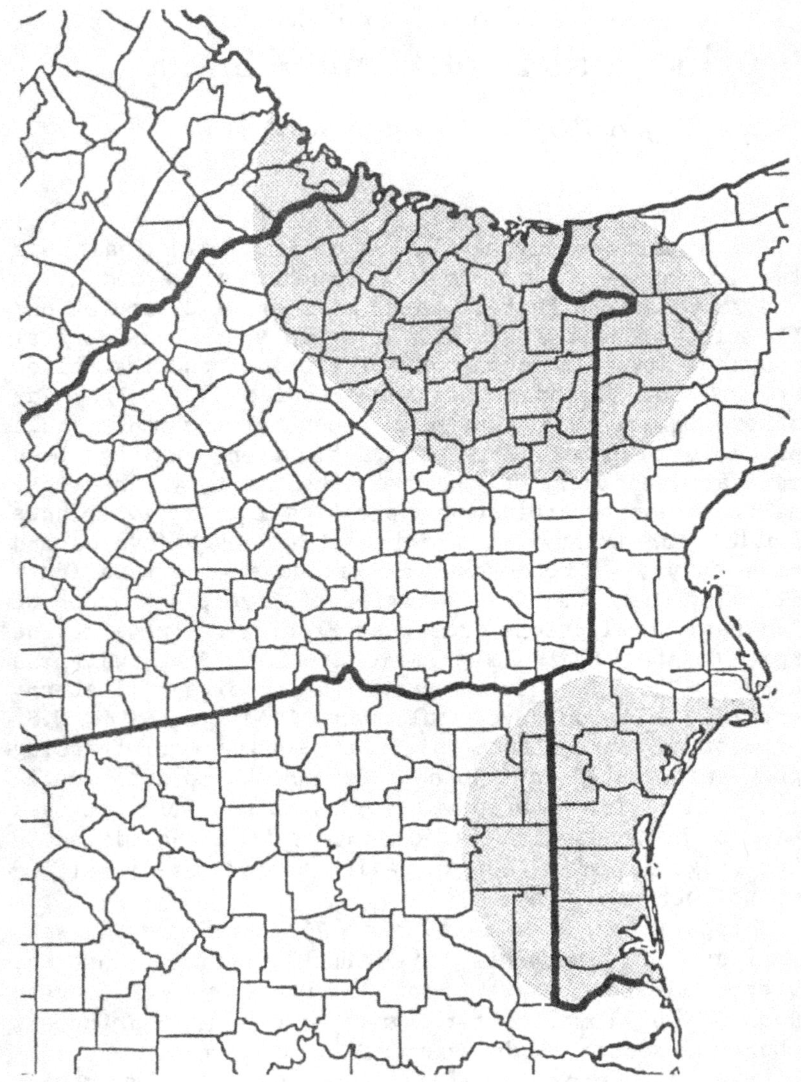

Figure 1. Native range of rabbiteye blueberry, _Vaccinium ashei_ Reade, 1985.

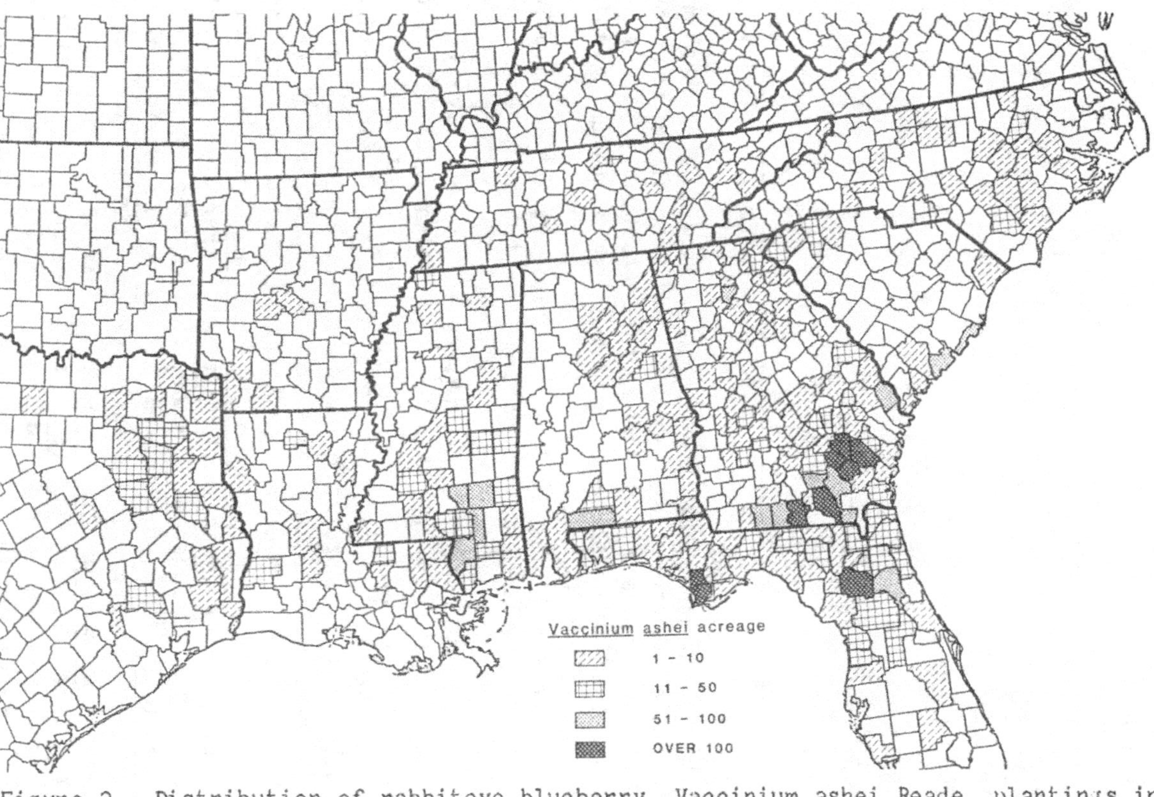

Figure 2. Distribution of rabbiteye blueberry, _Vaccinium ashei_ Reade, plantings in the United States, 1986.

Table 1. Rabbiteye blueberry acreage by state.

Georgia	2,513
Florida	885
Mississippi	624
Texas	246
South Carolina	178
Alabama	254
North Carolina	156
Louisiana	125
Tennessee	79
Arkansas	25
TOTAL	5,085

rabbiteye blueberries for Georgia, Florida, or Alabama. Rabbiteye blueberries are popularly perceived as being relatively disease and insect free (Austin 1979). After 30 years of cultivation this conception may be changing as acreage increases.

A survey (1985-1986) of entomologists in the leading blueberry states was conducted to determine the more important blueberry pests (Table 2). This survey indicated that rabbiteye blueberries are relatively free of pests, which is in sharp contrast to the pest situation for highbush blueberries (_V_. corymbosum L.) in New Jersey, North Carolina and Massachusetts, and for lowbush blueberries (_V_. angustifolium Aiton) in Canada and Maine.

Leading blueberry researchers contacted (1985-1986) expect rabbiteye blueberries to develop pest problems from cherry fruitworm, Grapholita packardi Zeller; plum curculio, Conotrachelus nenuphar (Herbst); Japanese beetle, Popillia japonica Newman; sharpnosed leafhooppers, Scaphytopius spp.; gypsy moth, Lymantria dispar (L.); blueberry maggot, Rhagoletis mendax Curran; cranberry weevil, Anthonomus musculus Say; blueberry leafminers, Caloptilia sp., Parornix sp.; cranberry rootworm, Rhabdopterus picipes (Olivier); and Putnam scale, Diaspidiotus ancylus (Putnam). A survey conducted during the 1985 growing season in Georgia among 7 native Vaccinium species and managed and unmanaged rabbiteye blueberries, found blueberries to be subject to attack from a number of insect pests (Tables 3 and 4).

Rabbiteye blueberry producers are fortunate since most of the pest species are sporadic problems. A continuation of the survey into 1986 with concentration on native, managed, and unmanaged rabbiteye plantings (Table 4) showed that only the cranberry fruitworm, <u>Acrobasis vaccinii</u> Riley, was a consistent pest throughout the state of Georgia. Even then less than 15% of the acreage was treated for this pest (Horton et al. 1987).

Blueberry maggot is an important pest of highbush blueberries in Michigan and New Jersey growing areas (Table 2). In Georgia blueberry maggots have been collected from <u>V</u>. <u>stamineum</u> L. (deerberry) and <u>V</u>. <u>arboreum</u> Marshall (sparkleberry), but maggots have not been collected from <u>V</u>. <u>ashei</u> (Tables 3 and 4). Interestingly these infested plants were growing adjacent to rabbiteye blueberry plantings that had fruit available for oviposition. In North Carolina, blueberry maggot populations tend to be discrete and localized, and often occur annually in the same fields (Meyer 1985). Milholland and Meyer (1984) note that in North Carolina blueberry maggots are generally found on sites where soil organic matter is high and moisture retention is good. Later ripening highbush cultivars and rabbiteyes are said to be more at risk. To date Georgia rabbiteye plantings have avoided maggot infestation.

Sharpnosed leafhopper, <u>Scaphytopius</u> <u>magdalensis</u> (Provancher), the vector of stunt disease, was collected from 3 native <u>Vaccinium</u> species and unmanaged rabbiteye blueberries in 1985 (Tables 3 and 4); however, no symptoms of stunt disease were observed on any plants. According to Dale and Mainland (1981) it may be advisable to periodically examine rabbiteye blueberry plantings, since the extent of infection that could develop is dependent upon the amount of infection present in both cultivated and wild species in the area, and upon the prevalence of the leafhopper vector.

The aphid, <u>Illinoia</u> <u>pepperi</u> (MacGillivray), the vector of shoestring virus in highbush and lowbush blueberries (Hancock et al. 1986) was collected in Georgia from <u>V</u>. <u>stamineum</u>; however, it was not collected from adjacent <u>V</u>. <u>ashei</u> cultivars (Tifblue, Woodard, Climax, Delite). To our knowledge shoestring virus has not been identified from any <u>Vaccinium</u> species in Georgia or Florida.

Blueberry bud mite, <u>Acalitus</u> <u>vaccinii</u> (Keifer), infestations were noted on several native <u>Vaccinium</u> species in south Georgia and north Florida in 1986, and

Table 2. Importance of blueberry insects in the various growing areas.

Insect	N.J.	Mich.	Me.	N.C.	Mass.	Canada	Ga.	Wash.
Blueberry Maggot Rhagoletis mendax	1	1	1	2	2	1	4	-
Cranberry Fruitworm Acrobasis vaccinii	2	1	4	1	2	4	2	-
Cherry Fruitworm Grapholita packardi	2	2	4	2	2-3	2	4	1
Plum Curculio Conotrachelus nenuphar	3	3	4	1	3	4	4	-
Putnam Scale Diaspidiotus ancylus	2-3	3	4	4	3	4	4	-
Other Scales	3	3	4	3	3	4	3	3
Cranberry Weevil Anthonomus musculus	2	4	4	3	2	4	4	-
Blueberry Bud Mite Acalitus vaccinii	3	4	4	2	3	4	4	-

Sharpnosed Leafhopper Scaphytopius spp.	1-2	3	4	1	3	4	4	-
Cranberry Rootworm Rhabdopterus picipes	3	4	4	3	4	4	4	-
Scarabeid Root Grubs Phyllophaga spp.	3	4	4	3	3	4	4	-
Crown Girdler Cryptorhynchus obliquus	3	4	4	4	4	4	4	-
Black Army Cutworm Actebia fennica	4	3	3	3	3	3	4	-
Spanworms Several species	4	3	2	4	2	2	3	3
Blueberry Thrips Frankliniella vaccinii	4	4	2	4	4	2	4	-
Blueberry Flea Beetle Altica sylvia	4	4	2	4	3	2	4	-
Blueberry Tip Borer Hendecaneura shawiana	4	3	-	2-4	3	4	-	-
Blueberry Leafminers Several Species	2-3	-	4	3	-	4	4	-

Table 2 (continued). Importance of blueberry insects in the various growing areas.

Insect	N.J.	Mich.	Me.	N.C.	Mass.	Canada	Ga.	Wash.
Termites Reticulitermes flavipes	4	4	4	3	4	4	3	–
Prionus Larvae Prionus spp.	–	4	–	4	3	4	4	–
Gypsy Moth Lymantria dispar	2-3	3	3	4	2	4	4	–
Leafrollers Several species	2	2	3	3	2-3	2	3	–
Stem Borer Oberea myops	4	4	–	3	3	4	3	–
Black Vine Weevil Otiorhynchus sulcatus	–	4	4	4	4	4	–	3
Orange Tortrix Argyrotaenia citrana	–	4	4	–	4	4	–	3
Tent Caterpillars Malacosoma spp.	4	3	3	3	3	2	–	3

Fall Webworm Hyphantria cunea	3-4	3	4	3	3	2	-	-
Japanese Beetle Popillia japonica	3	3	4	2-3	2	4	3	-
Aphids	2-3	2	-	4	3	4	4	2
Bud Gall Dasineura sp.	3-4	-	-	4	-	4	-	-
Blueberry Tip Midge Contarinia vaccinii	3-4	4	3	-	-	3	-	-

Code for insect rating:

1 Needs treatment virtually every year.
2 Almost always present but needs treatment only occasionally.
3 Rarely present in pest proportions.
4 Never present in pest proportions.
- Not found on blueberries

Source: Ratings made through correspondence with the following: J. Hollett, Canada-lowbush; M. E. Austin, Georgia-rabbiteye; H. Y. Forsythe, Maine-lowbush; C. R. Brodel, Massachusetts-highbush; J. W. Nelson, Michigan-highbush; J. R. Meyer, North Carolina-highbush; P. E. Marucci, New Jersey-highbush; and C. E. Shanks, Washington-highbush.

Table 3. 1985 arthropod survey of native Florida and Georgia _Vaccinium_ spp. (10 sweeps per sample, 3/26-10/11/85).

Arthropod	stamineum native	arboreum native	elliottii native	simulatum native	myrsinites native	corymbosum native
Leptoglossus phyllopus		*	*			
Banasa dimidiata	*		**		*	*
Euschistus servus						*
Misc. Pentatomidae			*		*	
Scaphytopius magdalensis	**	*			*	
Scaphytopius sp.	**		*		**	*
Acrobasis vaccinii						*
Lepidoptera Larvae	**	**	**	**	**	*
Tip Midges						

Rhagoletis mendax	**	***				
Neochlamisus sp.	*					
Coccinellidae	*	**		*		
Misc. Coleoptera	**	***	**		*	
Anthonomus sp.	*		*			
Chrysomelidae	*		*	*		**
Scale					**	
Thrips	***	***	**		*	***
Crickets/ Grasshoppers	**		**			
Aphids	***	**	**	**		**
Midges/Flies	**	***	***	**	**	
Ants	**	**	**	*	*	**
Spiders	***	***	***	*	***	**

Table 3 (continued). 1985 arthropod survey of native Florida and Georgia
Vaccinium spp. (10 sweeps per sample, 3/26-10/11/85).

Arthropod	stamineum native	arboreum native	elliottii native	simulatum native	myrsinites native	corymbosum native
Harvestmen	*	*	*			

* - rare - some present but average per sample no greater than 1.
** - moderate - average per sample greater than 1 but less than 5.
*** - numerous - average per sample 5 or greater.

even collected from managed rabbiteye blueberries. The bud mite infestations on managed rabbiteye were not worthy of treatment. Previous workers (Keifer 1941, Neunzig and Galletta 1977) have noted that rabbiteye blueberry had lower bud mite infestations than highbush blueberry but offered no plausible reasons for differences in susceptibility. Among highbush clones in cultivar breeding programs, major, consistent clonal differences in bud mite infestation are well known, and breeding for bud mite resistance is important in the highbush breeding programs in both North Carolina and Florida. (J. R. Ballington and P. M. Lyrene, personal communication).

The cherry fruitworm is an important pest of blueberries in Canada, Michigan, New Jersey, North Carolina and Washington (Table 2) and has been reported from apple, crabapple, and peach in Georgia (APHIS National Agricultural Pest Information System). The cranberry rootworm is an occasional pest of blueberries in the coastal lowlands of New Jersey and North Carolina (Table 2). It has been reported in Georgia but host records are unknown (APHIS National Agricultural Pest Information System). We were unable to collect either of these insects during our 1985-86 blueberry survey.

It is too early to predict possible economic losses from Japanese beetle, cranberry weevil, and gypsy moth, since these 3 pests are seldom widespread in the rabbiteye blueberry growing areas of the South. Pick-your-own producers have had problems with: Japanese beetles; yellowjackets, _Vespula maculifrons_ (Buysson); and red imported fire ants, _Solenopsis invicta_ Buren. These insects prefer the overripe fruit. Furthermore, their presence in an orchard may be a nuisance to pick-your-own customers.

Ants and spiders were the dominant arthropods collected in our sweepnet samples and represented 40% of the specimens from both managed and unmanaged rabbiteye blueberry plantings. According to Johnson et al. (1981), spider populations were higher and somewhat more diverse in wild blueberries than in commercial blueberries in northwest Arkansas. This was not readily apparent in our survey; however, we primarily sampled orchards which had received no insecticide or fungicide applications.

During our 1985-1986 survey of native _Vaccinium_ and cultivated rabbiteye blueberries, we collected 151 species of insects and mites, 82 of which fed on rabbiteye blueberries (Table 5). Many of these have been identified from blueberries in other states and have already been

Table 4.   1985 arthropod survey of native and cultivated Florida and Georgia <u>Vaccinium ashei</u>. (10 sweeps per sample 3/26-10/11/85).

	ashei native	ashei managed	ashei unmanaged
<u>Leptoglossus</u> <u>phyllopus</u>	*	*	**
<u>Banasa</u> <u>dimidiata</u>	*	*	**
<u>Euschistus</u> <u>servus</u>		*	
Misc. Pentatomidae	*		
<u>Scaphytopius</u> <u>magdalensis</u>			**
<u>Scaphytopius</u> sp.			*
<u>Acrobasis</u> <u>vaccinii</u>		**	
Lepidoptera Larvae	**	*	**
Tip Midges		***	
<u>Rhagoletis</u> <u>mendax</u>			
<u>Neochlamisus</u> sp.			
Coccinellidae		*	*
Misc. Coleoptera	*	*	*
<u>Anthonomus</u> sp.		*	*
Chrysomelidae	*	*	*
Scale	*	*	
Thrips	***	***	**

Crickets/ Grasshoppers		*	*
Aphids	*	*	*
Midges/Flies	*	**	
Ants	**	*	*
Spiders	***	***	***
Harvestmen			**

* - rare - some present but average per sample no greater than 1.

** - moderate - average per sample greater than 1 but less than 5.

*** - numerous - average per sample 5 or greater.

implicated as causing damage. Rabbiteye blueberries, even in their short domestication, already have experienced damage from cranberry fruitworm, Japanese beetle, leafrollers, leaftiers, <u>Oberea</u> stem borer, scales and fire ants (Horton et al. 1986). Pesticides occasionally are applied for control of these pests. We feel that consistent insecticide use, particularly if applied in a prophylactic fashion, will eventually cause problems if rabbiteye blueberries respond like many other sprayed crops.

The evolution of insect problems on cultivated crops, and in this century the attendant adaptation of these pests to insecticides, is well documented. Marucci (1977,1979) has described the development of insect problems for cultivated blueberries from his experiences with insect control in highbush blueberries in New Jersey (Table 6). Commercial concentrations of rabbiteye blueberries are presently planted where wild blueberries are indigenous and abundant. This surrounding reservoir of wild blueberries is likely to provide a source of pests and their natural enemies whose populations are balanced with one another. The monoculture system favors movement of native insects into cultivated plantings. The high expectations of modern agriculture, particularly in

114

Table 5. List of 80 arthropod pests field collected
and/or lab reared from rabbiteye blueberry in Georgia and
Florida during 1985-86.

COLEOPTERA
  CERAMBYCIDAE
    Oberea myops Haldeman
    Oberea tripunctata (Swederus)
  CHRYSOMELIDAE
    Bassareus brunnipes (Oliver)
    Bassareus detritus (Oliver)
    Neochlamisus gibbosus (Fabricius)
    Neochlamisus bimaculatus Karren
  CURCULIONIDAE
    Anthonomus musculus Say
    Anthonomus signatus Say
    Graphognathus leucoloma (Boheman)
    Graphognathus peregrinus (Buchanan)
    Pantomorus cervinus (Boheman)
  MELOIDAE
    Lytta aenea (Say)
  SCARABAEIDAE
    Euphoria sepuchralis (Fabricius)
    Macrodactylus angustatus Beauvois
    Popillia japonica Newman
DIPTERA
  CECIDOMYIIDAE
    Dasineura oxycoccana Johnson
HEMPITERA
  COREIDAE
    Acanthocephala terminalis (Dallas)
    Leptoglossus oppositus (Say)
    Leptoglossus phyllopus (Linnaeus)
  PENTATOMIDAE
    Banasa dimidiata (Say)
    Euschistus servus (Say)
    Euschistus tristigmus (Say)
    Thyanta accerra McAtee
    Nezara viridula (Linnaeus)
HOMOPTERA
  ALEYRODIDAE
    Aleyroplatus vaccinii Russell
    Tetraleurodes ursorum (Cockerell)
  APHIDAE
    Aphis gossypii Glover

CICADELLIDAE
  Scaphytopius magdalensis (Provancher)
COCCIDAE
  Mesolecanium nigrofasciatum (Pergande)
  Parthenolecanium corni (Bouche)
DIASPIDAE
  Abgrallaspis howardi (Cockerell)
FLATIDAE
  Anormenis septentrionalis (Spinola)
HYMENOPTERA
  TENTHREDINIDAE
    Neopareophora sp.
  FORMICIDAE
    Solenopsis invicta Buren
  VESPIDAE
    Polistes spp.
    Vespula maculifrons (Buysson)
LEPIDOPTERA
  ARCTIIDAE
    Estigmene acraea (Drury)
  GELECHIIDAE
    Aroga trialbamaculella (Chambers)
    Dichomeris vacciniella Busck
    Trichotaphe sp.
  GEOMETRIDAE
    Anacamptodes defectaria (Guenee)
    Euchlaena sp.
    Eutrapela clemataria (J. E. Smith)
    Hypagyrtis unipunctata (Haworth)
    Iridopsis larvaria (Guenee)
    Melanolophia canadaria choctawae Rindge
    Synchlora aerata (F.)
    Synchlora frondaria Guenee
  GRACILLARIIDAE
    Caloptilia vacciniella (Ely)
    Parornix sp.
  LASIOCAMPIDAE
    Malacosoma americanum (Fabricius)
  LIMACODIDAE
    Phobetron pithecium (J. E. Smith)
    Prolimacodes badia (Hubner)
  LYCAENIDAE
    Strymon melinus Hubner
  LYMANTRIIDAE
    Orgyia leucostigma (J. E. Smith)
  NOCTUIDAE
    Acronicta tritona (Hubner)

    Amphipyra pyramidoides Guenee
    Spodoptera frugiperda (J. E. Smith)
    Spodoptera latifascia (Walker)
  NOTODONTIDAE
    Datana major Grote & Robinsin
    Datana ministra (Drury)
    Datana sp.
    Schizura ipomoeae Doubleday
  PSYCHIDAE
    Thyridopteryx ephemeraeformis (Haworth)
  PYRALIDAE
    Acrobasis vaccinii Riley
  SATURNIIDAE
    Antheraea polyphemus (Cramer)
    Eacles imperialis (Drury)
  TORTRICIDAE
    Acleris sp.
    Archips rosana (Linnaeus)
    Argyrotaenia velutinana (Walker)
    Choristoneura obsoleana (Walker)
    Choristoneura rosaceana (Harris)
    Croesia curvalana (Kearfott)
    Epinotia sp.
    Pandemis limitata (Robinson)
    Platynota flavedana Clemens
    Pseudexentera sp.
    Sparganothis reticulatana (Clemens)
THYSANOPTERA
  THRIPIDAE
    Frankliniella bispinosa (Morgan)
    Frankliniella tritici (Fitch)
ACARI
  ERIOPHYIDAE
    Acalitus vaccinii (Keiffer)

---

cultural systems where mechanical harvesting is utilized,
demand high yields of consistent, blemish-free produce.
This places a great emphasis on pesticides to facilitate
effective exploitation of each crop's maximum potential.
Rabbiteye blueberry producers are already experiencing
pressure to spray since previously innocuous pests such as
ants, caterpillars, Japanese beetles, and leaffooted bugs
are harvested with the berries and may enter the marketing
channels unless efficiently cleaned. Growers in the
southeast have been reluctant to spray because of cost and

Table 6.   Evolution of insect problems in highbush blueberries.

---

1.   Blueberries planted in areas where wild blueberries and huckleberries were indigenous and abundant.
2.   Blueberry monoculture systems favorable for pest population buildup from native insects.
3.   Mechanical harvesting caused intensification of spray program.
4.   Marketing regulatory agencies imposed a standard of perfection to maintain high quality image.
5.   Aphids, scales, and leafminers became pests in intensely sprayed blueberry fields.
6.   Malathion no longer effective against leafrollers, leafminers, fall webworm, and Japanese beetle.
7.   Switch to Guthion destroyed natural enemies of aphid and now aphids need controlling.

---

secondarily because action thresholds do not exist for a number of pests. Blueberry growers are aware of the pesticide failure and insect resistance with other crops in the southeast and may be reluctant to spray, thus avoiding the scenario outlined by Marucci (1977, 1979).

Marucci (1979) first described the consequences of intense use of insecticides on highbush blueberry insects (Table 7). In all cases with the full spray program there were more blueberry aphids, blueberry leafminers and redbanded leafrollers than in the abandoned or wild (unsprayed) programs. Presently, only leafrollers among these three have been reported as pests in some managed rabbiteye blueberry orchards in the southeast. Circumstantial evidence is strong that cranberry fruitworm sprays have contributed to the leafroller problems. It could be said that pest control in blueberries is in the exploitative phase of crop protection (Smith 1969). Insecticide applications currently achieve effective control of insect pests on rabbiteye blueberry, but elevation of the pest status of secondary and occasional pests, pest resurgence, and resistance appear quite likely to develop if prophylactic spraying is practiced.

Rabbiteye blueberries currently enjoy an enviable position. With only cranberry fruitworm as a consistent

Table 7. Comparison of insect populations in wild, abandoned and cultivated highbush blueberry fields in New Jersey - 1979.

Type Field	Spray Program	Blueberry Aphids No./ Leaf Shoot	Blueberry Leafminer % Leaves Infested	Redbanded Leafroller % Leaves Damaged
Culti- vated	Full	23	29	17.5
Culti- vated	Full	13	12	4.9
Culti- vated	Full	10	9	1.3
Aban- doned	None	2.5	3.5	0
Wild	None	0.7	1.5	0

pest, our growers do not yet need to spray preventatively. The threat of insecticide resistance and the occurrence of insects in rabbiteye blueberries, which are already serious pests in highbush or lowbush blueberry production areas, mandates the development of better pest management strategies, i.e. integrated pest management. We presently have 14 lowbush and highbush blueberry pests for which cultural methods provide slight to very good control (Doehlert and Tomlinson 1947, Driggers 1927, Marucci 1977, 1979, Milholland and Meyer 1984, Scott et al. 1973, Sorenson 1982) (Table 8). Growers may have to rely on these methods in the future if pesticide restrictions prevent reentry of fruit pickers and expose endangered species and blueberry pollinators to residues. The pollination aspect is especially critical with rabbiteye blueberries since native bees are our most dependable pollinators (Cane 1987).

In our opinion, the biggest handicap to pest control is the lack of action thresholds for determining what pest

Table 8. Cultural methods of blueberry insect control.

Insect	Practice	Degree of Control
Putnam Scale	Prune old canes	+
Blueberry Bud Mite	Prune old canes	+++
Stem Gall	Prune out galls	++++
Stem Borer	Prune out borer	++
Tip Borer	Prune out borer	++++
Sawflies	Clean cultivation	++++
Cutworms	Clean cultivation	++++
Cranberry rootworm	Clean cultivation	+++
Cranberry fruitworm	Clean cultivation	+
Sharpnosed leafhopper	Clean cultivation and burning along woods	+
Plum curculio	Clean cultivation and burning along woods	++
Cranberry weevil	Clean cultivation and burning along woods	+++
Cherry fruitworm	Prune out and burn old canes	+
Blueberry maggot	Clean harvest and prompt picking	+

+ = Slight; ++ = Light to moderate; +++ = Good; ++++ = Very Good

population level needs a preventative treatment and when to treat. Action thresholds only exist for 6 blueberry pests, 3 for New Jersey (Marucci 1979), 1 for Michigan (Whalon 1983), 1 for North Carolina (Milholland and Meyer 1984) and 1 for Georgia (Horton et al. 1987) (Table 9). Currently labelled blueberry insecticides are effective in most cases. This makes use and perhaps overuse of insecticides an alluring option. Until adequate action thresholds are developed we continue to unnecessarily run the risk of inadvertantly setting the stage for pest resurgence, elevation of non-pests to pest status, and resistance.

The rabbiteye blueberry industry is in its infancy when compared to other fruit crops. At present we are fortunate to have very few pests, and if a

120

Table 9. Action thresholds for blueberry pests.

Pest	Time to make Population Count	Action Threshold
Fruittree leafroller	In winter or spring before egg hatch	50 eggs per 1,000 linear inch of twigs or - 250 eggs per 30 min. time count
Cranberry weevil	In spring when early variety blossoms show white	1 weevil puncture per 5 clusters or - 1 puncture per 40 flowers
Blueberry bud mite	In early autumn	When 25% of buds contain a total of 50 mites plus eggs
Aphid *Illinoia pepperi*	Late May-June	Average population of 36-100 aphids per bush or 7-15% of the terminals infested
Blueberry maggot	Late May-June	3 adults/baited yellow sticky board trap/week
Cranberry fruitworm	In spring during late bloom	When 1 bush in 5 has infested clusters

meticulous and judicious approach is taken we can develop the pest management tools to effectively manage our rabbiteye bluebery pests (Table 10). The opportunity exists to develop selective pest management strategies that allow for optimum production and, as much as possible, avoid the onset of prophylactic pesticide use.

Table 10. Simplified Georgia rabbiteye blueberry insect control chart.

Insect	Time to Treat	Insecticide (a.i.)
Cranberry Fruitworm	Check fruit from late bloom till 4 wks. after petal fall. Spray when 1 bush in 5 has infested clusters	Malathion (1 lb/A) Azinphosmethyl (0.5-0.75 lbs/A)
Scales	Before bud break	2% Superior Oil
Leafrollers, Leaftiers & Spanworms	When damaging infestation occurs	Azinphosmethyl (0.5-0.75 lbs/A) Malathion (1 lb/A)
Japanese beetle	When beetles appear and repeat at 7-day intervals as needed	Carbaryl (1.5-2 lbs/A)
Stem borer	In summer, prune out infested canes or wilted terminals	None
Fire ants	When ants are abundant enough to annoy customers (pick-your-own operations).Treat with a mound drench, using at least 1 gallon of finished mix per mound.	Diazinon (0.5 lb/ 100 gal)

ACKNOWLEDGEMENT

We are especially indebted to Philip E. Marucci and John E. Meyer for freely sharing their knowledge, experience, and unpublished data on blueberry pest management and to Paul Lyrene, Arlen Draper, Jim Ballington and Max Austin for sharing their knowledge of <u>Vaccinium</u> distribution, blueberry culture, and especially their views on the probable development of a rabbiteye blueberry pest complex.

We greatly acknowledge aid in identification of specimens by the following taxonomists from the: Biosystematic Research Institute, National Identification Service, Canada - P. T. Dang; Auburn Univ. - J. H. Cane, M. L. Williams; Univ. of Georgia - Ramona Beshear, J. O. Howell; Univ. of Florida - A. B. Hamon; Illinois Natural History Survey - W. E. LaBerge; Univ. of Illinois - S. H. Berlocher; Kansas State Univ. - R. W. Brooks; Univ. of Maryland - J. A. Davidson; Univ. of Missouri - M. S. Arduser; North Carolina State Univ. - H. H. Neunzig; Texas A & M Univ. - H. R. Burke; Utah State Univ. - J. B. Karren; Insect Identification and Beneficial Insect Introduction Institute, Agric. Res. Serv., USDA - D. M. Anderson, E. W. Baker, S. W. T. Batra, D. R. Davis, D. C. Ferguson, R. J. Gagne, R. D. Gordon, T. J. Henry, R. W. Hodges, J. P. Kramer, R. W. Pool, R. K. Robbins, D. R. Smith, M. B. Stoetzel, D. M. Weisman, R. E. White, D. R. Whitehead.

The authors are grateful to the following for input and review of our estimates of blueberry acreage: Alabama - Arlie Powell; Arkansas - Paul Horner; Florida - Tim Crocker; Georgia - Gene Hubbard; Louisiana and Mississippi - John Braswell; North Carolina - Mike Mainland; South Carolina - John Ridley; Tennessee - Alvin Rutledge; Texas - Terry Menges.

LITERATURE CITED

Austin, M. E. 1979. Rabbiteye blueberries. Fruit
    Varieties Journal 33(2):51-53.
Brightwell, W. T. and O. Woodard. 1955. Observations on
    breeding blueberries for the southeast. Proc. Am. Soc.
    Hort. Sci. 65:274-278.
Cane, J. 1987. The importance of pollination.
    Proceedings of 3rd Biennial Southeast Blueberry
    Conference and Trade Show. pp. 42-49.

Dale, J. L. and C. M. Mainland. 1981. Stunt disease in
rabbiteye blueberry. HortScience 16(3):313-314.

Doehlert, C. A. and Tomlinson, W. E., Jr. 1947. Blossom
weevil on cultivated blueberries. N. J. Agr. Exp. Sta.
Circ. 504, 8 pp.

Driggers, B. F. 1927. Notes on the life history and
habits of the blueberry stem borer, _Oberea_ _myops_ Hald.
on cultivated blueberries. J. NY Entomol. Soc.
37:67-73.

Galletta, G. J. 1975. Blueberries and cranberries, pp.
154-196. _In_ J. Janick and J. N. Moore (eds.), Advances
in fruit breeding, Purdue Univ. Press, West Lafayette,
IN.

Hancock, J. F., K. M. Morimoto, M. P. Pritts, and D. C.
Ramsdell. 1986. Blueberry shoestring virus in natural
populations of highbush and lowbush blueberry.
HortScience 21(4):1059-1060.

Horton, D. L., H C Ellis, and P. Bertrand. 1986.
Blueberry spray guide, GA. Coop. Ext. Serv. Bull., 4 pp.

Horton, D., J. Payne, A. Amis, and D. Warnock. 1987.
Insect management in Georgia rabbiteye blueberries.
Proc. 3rd Biennial Southeast Blueberry Conference and
Trade Show, pp. 66-71.

Johnson, D. L., L. C. Hopkins, and J. S. Heiss. 1981.
Arthropods associated with wild and cultivated
blueberries in NW Arkansas, Arkansas Farm Research.
30(5):4.

Keifer, H. H. 1941. Eriophyid studies XI. Bull. Calif.
Dept. Agr. 30:196-204.

Lyrene, P. M. and W. B. Sherman. 1979. The rabbiteye
blueberry industry in Florida - 1887 to 1930 - with
notes on the current status of abandoned plantations.
Economic Botany 33(2):237-243.

Marucci, P. E. 1966. Insects and their control. pp.
199-235. _In_ P. Eck and N. F. Childers (eds.), Blueberry
Culture, Rutgers Univ. Press, New Brunswick, NJ.

Marucci, P. E. 1977. The control of blueberry insects in
New Jersey. Acta Hort. 61:187-196.

Marucci, P. E. 1979. The control of blueberry insects
and blueberry pollination in New Jersey. Proceedings of
the 4th North American Blueberry Research Workers
Conference pp. 197-202.

Meyer, J. R. 1985. What's bugging your blueberries?
Proc. North Amer. Blueberry Council Annual Meeting
20:68-73.

Milholland, R. D., and Meyer, J. R. 1984. Diseases and
    arthropod pests of blueberries. N. C. Agr. Res. Ser.
    Bull. 468, 33 pp.
Neunzig, H. H. and G. J. Galletta. 1977. Abundance of
    the blueberry bud mite (Acarina:Eriophyidae) on various
    species of blueberry. J. Ga. Entomol. Soc.
    12(2):182-184.
Phipps, C. R. 1930. Blueberry and huckleberry insects.
    Maine Agr. Exp. Sta., Bull. 356, 232 pp.
Scott, D. H., Draper, A. D., and Darrow, G. M. 1973.
    Commercial blueberry growing. USDA Farmer's Bull. No.
    2254, 30 pp.
Smith, R. F. 1969. The new and old in pest control.
    Accad. Naz. dei. Lincei, Rome. 366:21-30.
Sorenson, K. A. 1982. Some factors affecting insect and
    mite infestations on blueberry. N. C. Agr. Ext. Ser.
    Insect Note 3-B (Revised).
Whalon, M. E. 1983. _Illinoia pepperi_ (MacG), the
    blueberry shoestring virus vector. Proc. of the North
    American Blueberry Workers Conference pp. 48-65.

6

Entomology of Johnsongrass/ Sorghum/Sorghum Midge and Agriculture

G. L. Teetes

Sorghum midge, <u>Contarinia</u> <u>sorghicola</u> (Coquillett) (Diptera: Cecidomyiidae), and host grasses of the genus <u>Sorghum</u> constitute a naturalized system in North America of seemingly permanent success that directly affects agriculture. Of interest to the topic of this symposium is the role of johnsongrass, <u>Sorghum</u> <u>halepense</u> (L.) Pers., a persistent and troublesome weed, in the population dynamics of sorghum midge, a persistent and troublesome insect pest of cultivated "grain" sorghum, <u>Sorghum</u> <u>bicolor</u> (L.) Moench. The subject has management implications by cultural means and plant resistance. The relationship of sorghum, sorghum midge, and johnsongrass is examined in terms of biologies, origin and transport, population dynamics, and management.

BIOLOGIES

Sorghum

Sorghum is a large stemmed tropical grass having the ability to grow to great heights. The species is extremely diverse and present collections contain over 17,000 distinct cultivars (Miller 1980). Sorghum is a perennial, grown as annual, is a self-pollinating species, but is handled in advanced agricultural situations as a cross-pollinating crop. The diverse types in the species are all dipolids with a 2N=20 chromosome number, and they will intercross (Schertz 1980). There are related grassy species with 2N=40. Cytoplasmic genetic male sterility is extremely important to hybridization in sorghum. The cornerstone to much of the present crop improvement in sorghum is the imaginative Sorghum Conversion Program (Stephens et al. 1967). This joint Texas Agricultural Experiment Station - USDA, ARS project changes tall, late or non-flowering sorghums from the tropics into short, early forms which can be used in all areas of the world, but especially in temperate zones (Miller 1980). This is done by substituting up to eight genes which control height and maturity to obtain the desired genotypes. Sorghum originated near the equator and is sensitive to the length of day. Knowing the genetics of height and maturity, and the response to daylength, it is possible to use the facilities of the USDA's tropical research station at Mayaquez, Puerto Rico and a temperate selection site at Chillicothe, Texas to make vast amounts of germplasm available for further plant improvement. Selections are made from the World Collection which are judged to offer the greatest diversity and eliteness. The original cross and four backcrosses of the exotic line onto the dwarf, nonphoto-sensitive line with selection in each generation allows the recovery of over 98% of the germplasm in each entry. During the last backcross, which is done by hand emascultation, the cross is made using the alien line as the female. This allows the recovery of the cytoplasm of the converted line. At present, there are 1,216 lines in the program.

Johnsongrass

Johnsongrass is a stout, erect, perennial grass that spreads by seeds and by long creeping rhizomes (Holm et al. 1977). Its range as a weed extends from lat 55°N to 45°S. There are many ecotypes of the species with 55 distinct vegetative types collected from many areas of the U.S.

Propagation and dissemination is principally by seeds. After maturity, the seeds shatter readily from the panicle. Seeds have remained viable for long periods, germinating after a dormant period. Johnsongrass is a very heavy seed producer; however, its superior ability to compete with other plants and its persistence in the face of the most intensive control measures surely result from the long, very vigorous, and highly adaptable rhizome-root system. It has been estimated that johnsongrass can produce, per hectare, 600 km of rhizomes weighing 33 metric tons. Individual plants may produce 5,000 nodes in one growing season. In general, the rhizome system is made up of primary, secondary and tertiary rhizomes. The primary structures are alive at the beginning of the growing season, providing buds for renewed growth. Extensions from the main rhizomes become the secondary structures which surface and give rise to new plants. Tertiary rhizomes, which grow out from the base of the plant at flowering time, are large, usually go deep into the soil, and usually continue to grow until the advent of cold or dry weather. These tertiary rhizomes produce new plants in the following season. At the beginning of the season, either axillary or terminal buds may develop into new plants with aerial stems which subsequently develop crowns and tillers. A high temperature is necessary for renewed activity of rhizome buds after a dormant period. Data are unavailable, but empirical evidence would indicate that heat unit requirements for initial johnsongrass development in the spring parallels that of sorghum midge diapause termination.

Sorghum Midge

The sorghum midge is considered to be the most serious insect pest of sorghum because of its wide distribution and the direct damage resulting in kernel loss (Young and Teetes 1977). The adult sorghum midge is a small fly, 2 mm long and orange in color (Harris 1969). A female midge deposits about 50 eggs between the glumes in flowering spikelets of sorghum. The immature stages develop cryptically within spikelets and larval feeding inhibits kernel formation. Depending on environmental conditions, a generation is completed in a range of 14 to 22 days (Walter 1941). At favorable temperatures, mean development time from egg to adult is 16 days allowing for numerous generations per season which accounts for the build-up of extremely high midge densities, especially when the sorghum flowering period is extended by successive planting dates. A single feeding larva is sufficient to destroy a kernel. When midge densities are high, all grain can be lost, and spikelets may contain more than one midge.

Adult male and female midges are short lived; emerging, mating, ovipositing and dying on the same day. Consequently, a new brood of midges occurs each day.

Adult midge emergence from infested spikelets and oviposition in flowering spikelets are influenced by time of day, temperature and moisture, and are reflected in hourly, daily and seasonal density fluctuations (Fisher and Teetes 1982, Fisher et al. 1982). Emergence is completed during the morning hours. Cool temperatures delay midge emergence. Males hover around the panicle from which they emerge and mate with females as soon as they emerge. After mating, females leave the panicle from which they emerged and disperse to flowering host panicles. Abundance of ovipositing females in flowering hosts is directly related to prior emergence patterns from infested panicles. The greatest number of ovipositing females in hot climates occurs at 1130, two hours after maximum female emergence from infested panicles, and abundance is least between 1830 and 0630. A variable proportion of sorghum midge larvae in each generation construct silken cocoons and enter a state of diapause within spikelets of the host plant (Baxendale and Teetes 1983a). Typically, these spikelets fall to the ground and become covered with litter, or are disked into the soil along with plant residues. Spring emergence times and yearly emergence distributions of overwintered sorghum midges are a function of soil temperature and moisture (Baxendale and Teetes 1983b). Adult midges initiate emergence after accumulating 431 centigrade heat units (based on mean daily 10cm soil temperatures starting 1 April) above a threshold temperature of 14.8°C, whereas 679 and 977 heat units are required for 50 and 95% emergence, respectively. The time that midges enter diapause one year has little effect on the time or distribution of emergence the following spring. Midges do not terminate diapause and emerge as adults during the same year they entered diapause. Almost 1/4 of the midges entering diapause during a season fail to emerge the subsequent spring and emerge the second spring, and almost 3% do not emerge until the third spring, but times and distribution of emergence are similar for any year.

ORIGINS

<u>Sorghum</u>

While there is consensus of opinion that <u>Sorghum bicolor</u> (L.) Moench, originated in Africa, there is less agreement concerning the region of Africa in which this species was first domesticated (Murdock 1959, De Wet and Huckabay 1967, Harlan 1975, Doggett 1976, Mann et al. 1983). Western Africa is considered by some to be the aboriginal home of cultivated sorghum, but most likely it is the northeast quadrant of Africa (Doggett 1970, Murdock 1959). Based on comparative morphological studies and using numerical taxonomy, three independent centers of sorghum domestication have been proposed - the Ethiopian area of eastern Africa, tropical western Africa, and southeastern Africa (De Wet and Huckabay 1967). Harlan (1975) believed sorghum typified a "noncentric crop"; that is, a crop whose current distribution of genetic variability revealed no evident center of origin or center of diversity. After considering the distribution of races of sorghum in Africa, Harlan (1975) concluded that initial domestication of sorghum occurred in a long belt across central Africa. Doggett (1976) reviewed the findings of Harlan and others regarding the origin of sorghum cultivation and concluded that sorghum was first cultivated by the Cushites in the Ethiopian highlands of eastern Africa.

Following the initial domestication of sorghum around 3000 to 4000 B.C., different races were domesticated by various tribes in different parts of Africa as knowledge of crop cultivation spread. During the period that the Durra race was diversified in the Ethiopian-Sudan region, the Guinea race was developed in west Africa, at least in part by Mande tribesmen. The Bantu people, who migrated from the Cameroons and cultivated Malaysian food plants such as bananas and coco yams as they moved along the southern edge of the Congo forests, later adopted sorghum as a staple food as they began to populate the dry savannas east of the forest belt. From Tanzania southward the Kafir race of sorghum came to be associated with them, just as the Caudatum race became associated with the Nilotic and Nilo-Hamitic people of Uganda and western Kenya. The fact that sorghum is an extremely diverse species with much morphological variability resulted in its utilization in different ways by various tribes, from building materials for huts, furniture, and mats to beer, sugar, and cereal production (De Wet and Huckabay 1967).

From Ethiopia and Sudan, cultivation of the Durra race spread into the Near East and to India soon after 2000 B.C. Sorghum was probably not cultivated in ancient Egypt, and it is also apparently absent from the extensive archaeological excavations of early farming sites in the

Near East. In the first century (approximately 60-70 A.D.), Pliny mentioned the introduction of sorghum into Italy by caravans from India.

The spread of sorghum cultivation into Southeast Asia and China also occurred "around the beginning of the Christian era"; but, there are not authentic records of sorghum in China before the 13th century, despite several suggestions of an earlier arrival (Doggett 1970, Martin 1970). It was in Southeast Asia where the crop first came in contact with wild diploid forms of sorghum such as S. propinquum. Such genetic exchange between the wild and cultivated forms may have given rise to the characteristic koaliang types of China, Manchuria, and Japan (Doggett 1976).

Sorghum was probably introduced into the Western Hemisphere from western Africa, when "guinea corn" or guinea kafir (Quinby 1974), rural branching durra or white milo maize (Martin 1936) and possibly "chicken corn" or shattercane (Martin 1970) were introduced (Doggett 1976). Sorghum probably first entered the U.S. along with slaves, although there is apparently no record of such an introduction. Sorghum is believed to have been under cultivation in the U.S. prior to the end of the 18th century, as Benjamin Franklin is credited with introducing broomcorn prior to that time (Martin 1970). Guinea corn was also under cultivation in the U.S. by then (Martin 1936). The value of guinea corn was mentioned to the Philadelphia Agricultural Society in 1810, and its virtues extolled in South Carolina in 1882 (Quinby 1974). Shortly after the turn of the century, two species of sorghum, S. saccharatum and S. vulgare, were included in the "Manual of Botany for the Northern and Middle States of America" (Eaton 1922). The first of these species was referred to as broomcorn, while the common name of the latter was given as "indian millet" or "coffeecorn". The "var. bicolor" was listed as a subspecies of S. vulgare. Broomcorn was said to be introduced "from the East Indies", while the origin of S. vulgare was described simply as "exotic."

Johnsongrass

The tetraploid Halepensia group of sorghum arose as a segmental allotetraploid between two, 20-chromosome species (Doggett 1970). One species was from the Arundinacea subsection, and the other from a related form

with the rhizomatous character. It is unlikely that
johnsongrass, <u>S</u>. <u>halepense</u>, originated from <u>S</u>. <u>virgatum</u>, a
North African variety of the Arundinacea with rudimentary
rhizomes. Most likely, johnsongrass resulted from the
wild Arundinacea and <u>S</u>. <u>propinquum</u>, followed by chromosome
doubling. The area where johnsongrass arose could well
have been in the Indian region, where a few specimens of
both Arundinacea sorghums and diploid rhizomatous
Halepensia sorghums have been recorded. These assumptions
suggest that johnsongrass is not the progenitor of the
cultivated sorghums, but certain taxonomic and genetic
characters in common show a close relationship. Johnsongrass has been remarkably successful, and has spread into
a wide range of conditions.

Although today most sorghums are grown for grain production in the U.S., the first sorghums to receive widespread publicity and USDA research attention were those
intended for forage or sugar production. Probably the
first sorghum to receive widespread U.S. attention was
johnsongrass, introduced for forage production. Contra-
dictory accounts have been presented in the literature
regarding the time of its initial introduction into the
U.S. There is agreement among various authors that
johnsongrass was introduced into North America (Ball 1902,
Vinall 1926, McWhorter 1971). Most authorities attribute
the introduction of johnsongrass into the U.S. to an
employee of John H. Means, governor of South Carolina, who
brought johnsongrass seed with him from Turkey upon his
return to South Carolina around 1830. As the grass became
established, people referred to it as "Mean's grass."
Approximately 10 years later, Colonel William Johnson,
the owner of a large plantation at Marion Junction,
Alabama, visited South Carolina. On his return trip to
Alabama, he is said to have carried "a quantity of
johnsongrass seed", which he sowed on his farm in the
fertile bottom lands of the Alabama River. As the grass
became established, it supposedly received the name
johnsongrass. A second account suggested that there were
probably multiple introductions of johnsongrass, including
some prior to 1830 in the states of Mississippi and
Georgia (McWhorter 1971). References to johnsongrass after
1834, suggested that johnsongrass seed was imported from
Jamaica both prior to and after 1850 (USDA 1874).

At the turn of the century, johnsongrass seed was
being marketed in Alabama, Mississippi, Louisiana, and
Texas, and the species had spread throughout the southern

and southeastern U.S. from Texas to Florida (Pieters 1902, Ball 1902). By the beginning of the last decade, it had further extended its range "as far north as central New York, New Hampshire, and Vermont in the East" and to parts of Oregon and Washington in the West (USDA 1970). Although the species initially received a great deal of favorable publicity (USDA 1874, 1875), the rapid spread and persistence of the grass in areas of the South where it had been introduced became a matter of increasing concern (Vasey 1882). In 1895, the Texas legislature passed a law prohibiting the distribution or sale of johnsongrass seed, roots, or hay (Ball 1902). By the following year, USDA scientists were gathering data on johnsongrass as a weed (Coville 1896). In 1901, the USDA appropriations bill passed by Congress instructed that agency to prepare a report on johnsongrass (Ball 1902). Regional markets for the sale of johnsongrass hay were still in operation in Georgia, Alabama, Mississippi, and San Antonio, Waco, and Fort Worth, Texas, in the mid-1920's (Vinall 1926). However, the increasing availability of sudangrass and other less persistent and more nutritious forage grasses, as well as the conversion from animals to internal combustion engines to power farm machinery and transportation (Quinby 1974), probably contributed to an increasing tendency to regard johnsongrass as a noxious weed rather than as a desirable grass species (Talbot 1928, Hitchcock 1935, USDA 1970). Johnsongrass is now classified as one of the 10 worst weeds of field crops in the U.S. (Anderson 1969), both because of its exceptional ability to compete with field crops and available nutrients and moisture, and because it serves as an alternate host for several plant pathogens (Taylor and Pares 1968, Milbrath and Tosic 1974) and insect pests (Hobbs et al. 1978, Bottrell 1971) including the sorghum midge (Dean 1911). Although some earlier authors recommended the destruction of johnsongrass as a means of suppressing sorghum midge density, the role of this ubiquitous perennial in the population dynamics of the sorghum midge must be clearly established.

Sorghum Midge

The dates of first discovery of sorghum midge in various sorghum-growing countries might appear to suggest that, after its first discovery in the United States in

1895, it spread to Hawaii, to the West Indies, to Indonesia and Australia and to Sudan and that, from Sudan, it gradually spread throughout Africa. Doggett (1970) stated in his book <u>Sorghum</u> that there seems no reason to doubt that the sorghum midge is of New World origin, and a rather recent immigrant to Africa. While there are few prehistoric and historic facts about the origin of sorghum midge, there is certainly strong evidence to suggest that Doggett's opinion is incorrect (Harris 1963, Barnes 1956). There is ample support of the contention that sorghum midge originated with sorghum in its aboriginal home (Harris 1961, 1985). The support of this contention is as follows:

1. Most sources of plant resistance to sorghum midge are "Zera Zera" types from Ethiopia and Sudan;

2. Botanical examination of all relevant material of the genus <u>Sorghum</u> in the Kew Herbarium and the herbarium of the British Museum of Natural History revealed the presence of midge larvae and pupae in 45 samples which proved that sorghum midge was present in Sudan in 1869, 26 years before its discovery in the United States;

3. Only grasses of the genus <u>Sorghum</u> are hosts of the sorghum midge and no sorghums are indigenous to the United States;

4. Species diversity of <u>Contarinia</u> and perhaps associated parasites is likely greater in Africa than the United States;

5. The sorghum midge is far more severe a pest in the United States than in northeast Africa.

There seems no doubt that the sorghum midge is indigenous to Africa and an alien Texan and American.

POPULATION DYNAMICS

Fundamental to the description of the association of sorghum, johnsongrass and sorghum midge is the construction of a seasonal dynamics model. Sorghum midges overwinter subterraneously as diapausing larvae within spikelets of <u>Sorghum</u>. In the spring, adults emerge from the soil before cultivated sorghum in the area begins to flower. However, the time that johnsongrass flowers coincides perfectly with first adult spring emergence from diapausing larvae. The adults from diapausing larvae oviposit in spikelets of johnsongrass to produce the season's first generation. Johnsongrass constitutes the host for the first generation as well as a second generation that is produced before cultivated sorghum begins to flower. Once early planted sorghum begins to flower, most midges disperse to sorghum where economic densities can be reached in a single additional generation (Baxendale et al. 1984ab). Sorghum flowering after this time is subject to severe midge damage. Very late in the season, sorghum midge densities generally decline to non-economic levels. Despite the apparent preference for panicles of cultivated sorghum, midge infest johnsongrass throughout the season.

SIGNIFICANCE TO AGRICULTURE

The most effective method of reducing losses by sorghum midge presently available is avoidance by the cultural practice of uniform, regional planting of sorghum early in the growing season (Teetes 1985). However, such planting is not always possible. Planting periods may be delayed or extended due to drought or by frequent spring rains. Natural enemies of sorghum midge include most general predators found in a sorghum field and at least four species of hymenopterous parasites. However, over the course of a season only about 8% of sorghum midges infesting sorghum and 20% infesting johnsongrass are parasitized (Baxendale et al. 1983c). Midge density suppression by natural enemies appears to be inadequate to protect sorghum even though parasitism is perpetuated in johnsongrass.

When weather conditions at planting time result in staggered planting dates, the only control measure available to protect the later plantings has been the use of insecticides to kill females prior to oviposition. Current recommendations suggest that applications begin when 25-30% of the panicles in a field are at anthesis ("flowering" or "yellow bloom") and there is an average of one adult midge per panicle. Additional applications might be needed at three to five day intervals during the remainder of the flowering period in order to maintain adequate control. However, attempting to kill adult female midges with the contact insecticides now labeled for use in sorghum is difficult. The short life cycle of the sorghum midge, the occurrence of broods and overlapping generations, the flowering characteristics of sorghum and the relatively brief time span required for females to infest a field require that fields be scouted and sprayed frequently to control new broods of females throughout a bloom period that may last for two weeks or more.

The naturalized association between sorghum, johnsongrass and sorghum midge has significance beyond population dynamics and economic pest status. Evidence would strongly support a sympatric relationship that has important management implications. A major consideration is the relative importance of sorghum and johnsongrass as hosts of sorghum midge. Based on the accumulated knowledge of the population dynamics of sorghum midge, johnsongrass is by far the more important host. Our understanding of the system would lead us to believe that the pest could survive well in the absence of cultivated sorghum, but would likely perish, at least as an economically important key pest, without johnsongrass. So, in terms of survival, johnsongrass is the host for sorghum midge and sorghum is the alternate host.

From a more strictly management standpoint, the sympatric evolution of sorghum midge and sorghum would have strong implications for the presence of plant resistance (Teetes 1985). The long-term association of the two organisms should have yielded significant resistant germplasm. In fact, it has. The major sources of midge resistance being used in breeding for resistance programs today are "Zera Zera" sorghums, which are traceable to Ethiopia and Sudan (Johnson et al. 1973, Wuensche et al. 1981).

Johnsongrass is a host of many of the insect pests and diseases of sorghum. However, some benefit can be imagined from this relationship. One could be parasite maintenance. Another could be genetic stability of pathogens and insect pests such as sorghum midge, which might prevent biotype development that results from insecticide or resistant host selection. This puts a much different light on the recommendations to control johnsongrass. Besides, it doesn't seem that we can control the damn weed, so why not stress its value.

REFERENCES CITED

Anderson, L. E. 1969. The ten worst weeds of field crops: johnsongrass. Crops and Soils, 22(3):7-9.
Ball, C. R. 1902. Johnsongrass: report of investigations made during the season of 1901. U.S. Dept. Agric. Bur. Plant Ind. Bull. 11. 24 pp.

Barnes, H. F. 1956. Gall midge injurious to the sorghums. In: Gall Midges of Economic Importance. Barnes, H. R. London, Crosby Lockwood. Vol. VII. 151-181 pp.

Baxendale, F. P. and G. L. Teetes. 1983a. Factors influencing adult emergence from diapausing sorghum midge, Contarinia sorghicola (Diptera: Cecidomyiidae). Environ. Entomol. 12:1064-1067.

Baxendale, F. P. and G. L. Teetes. 1983b. Thermal requirements for emergence of overwintered sorghum midge (Diptera: Cecidomyiidae). Environ. Entomol. 12:1078-1082.

Baxendale, F. P., C. L. Lippincott and G. L. Teetes. 1983c. Biology and seasonal abundance of Hymenopterous parasitoids of sorghum midge (Diptera: Cecidomyiidae). Environ. Entomol. 12:871-877.

Baxendale, F. P., G. L. Teetes and P. J. H. Sharpe. 1984a. Temperature-dependent model for sorghum midge (Diptera: Cecidomyiidae) spring emergence. Environ. Entomol. 13:1566-1571.

Baxendale, F. P., G. L. Teetes, P. J. H. Sharpe, and H. Wu. 1984b. Temperature-dependent model for development of nondiapausing sorghum midge (Diptera: Cecidomyiidae). Environ. Entomol. 13:1572-1576.

Bottrell, D. G. 1971. Entomological advances in sorghum production, pp. 28-40. In: PR 2940. Grain sorghum research in Texas...1970. Tex. Agric. Exp. Stn. Consol. PR-2938-2949. 120 pp.

Coville, F. V. 1896. Weed investigations, pp. 96-7. In: Report of the Secretary of Agriculture. U.S. Gov. Print. Office, Washington. 269 pp.

Dean, W. H. 1911. The sorghum midge. U.S. Dept. Agric. Bur. Entomol. Bull. 85(IV):39-85.

De Wet, J. M. and J. P. Huckabay. 1967. The origin of Sorghum bicolor. II. Distribution and domestication. Evolution 21:787-802.

Doggett, H. 1970. Sorghum Longmans, Green and Co. Ltd., London.

Doggett, H. 1976. Sorghum, pp. 112-7. In: Evolution of Crop Plants. Ed. N. W. Simmonds. Longman, New York. 339 pp.

Eaton, A. 1922. Manual of botany for the northern and middle states of America, containing generic and specific descriptions of indigenous plants and common cultivated exotics, growing north of Virginia. Websters and Skinners, Albany, New York. 536 pp.

Fisher, R. W. and G. L. Teetes. 1982. Effects of moisture on sorghum midge (Diptera: Cecidomyiidae) emergence. Environ. Entomol. 11:946-948.

Fisher, R. W., G. L. Teetes and F. P. Baxendale. 1982. Effects of time of day and temperature on sorghum midge emergence and oviposition. Tex. Agric. Exp. Stn. PR-4029. 8 pp.

Harlan, J. R. 1975. Crops and man. Amer. Soc. Agron., Madison, Wisc. 295 pp.

Harris, K. M. 1961. The sorghum midge, Contarinia sorghicola (Coq.), in Nigeria. Bull. Entomol. Res. 52:129-46.

Harris, K. M. 1963. The sorghum midge complex (Diptera: Cecidomyiidae). Taxonomy, biology and assessments of field populations. Bull. Entomol. Res. 63 (Pt. 2):305-325.

Harris, K. M. 1969. The sorghum midge. World Crops. 21:176-9.

Harris, K. M. 1985. The sorghum midge: A review of published information, 1895-1983. Proc. International Sorghum Entomology Workshop. ICRISAT. 1985, pp. 15-21 July 1984. Texas A&M University, College Station, TX, USA. 229-232 pp.

Hitchcock, A. S. 1935. Manual of the grasses of the United States. U.S. Gov. Print. Office, Washington. 1040 pp.

Hobbs, J. R., G. L. Teetes, J. W. Johnson, and A. L. Wuensche. 1979. Management tactics for the sorghum webworm in sorghum. J. Econ. Entomol. 72:362-6.

Holm, L. G., D. L. Plucknett, J. V. Pancho and J. P. Herberger. 1977. Sorghum halepense (L.) Pers. In: The World's Worst Weeds. The University Press of Hawaii, Honolulu. 54-61 pp.

Johnson, J. W., D. T. Rosenow and G. L. Teetes. 1973. Resistance to the sorghum midge in converted exotic sorghum cultivars. Crop Sci. 13:754-755.

Mann, J. A., C. T. Kimber and F. R. Miller. 1983. The origin and early cultivation of sorghums in Africa. Tex. Agric. Exp. Stn. B-1454. 21 pp.

Martin, J. H. 1936. Sorghum improvement. In: U.S. Dept. Agric. Yearb. U.S. Gov. Print. Office, Washington. 789 pp.

Martin, J. H. 1970. Sorghum production and utilization. In: History and Classification of Sorghum. Eds. J. S. Wall and W. M. Ross. The Avi Publishing Company, Inc., Westport, Conn. 702 pp.

McWhorter, C. G. 1971. Introduction and spread of johnsongrass in the United States. Weed Sci. 19(5):496-500.

Milbrath, G. M. and M. Tosic. 1974. Occurrence of maize chlorotic dwarf virus in Illinois. Plant Dis. Rep. 58(5):420.

Miller, F. R. 1980. The breeding of sorghum. In: Biology and Breeding for Resistance to Arthropods and Pathogens in Agricultural Plants. Ed. M. K. Harris. Tex. Agric. Exp. Stn. MP-1451. 128-136 pp.

Murdock, G. P. 1959. Africa, its people and their culture history. McGraw-Hill, New York. 456 pp.

Pieters, A. J. 1902. Agricultural seeds -- where grown and how handled. In: U.S. Dept. Agric. Yearb. 1901. U. S. Gov. Print. Office, Washington. 846 pp.

Quinby, J. R. 1974. Sorghum improvement and the genetics of growth. Texas A&M University Press, College Station. 108 pp.

Schertz, K. F. 1980. Biology of sorghum. In: Biology and Breeding for Resistance to Arthropods and Pathogens in Agricultural Plants. Ed. M. K. Harris. Tex. Agric. Exp. Stn. MP-1451. 124-127 pp.

Stephens, J. C., F. R. Miller, and D. T. Rosenow. 1967. Conversion of alien sorghums to early combine genotypes. Crop Sci. 7:396.

Talbot, M. W. 1928. Johnsongrass as a weed. U.S. Dept. Agric. Bull. 1537. 10 pp.

Taylor, R. H. and R. D. Pares. 1968. The relationship between sugarcane mosaic virus and mosaic viruses of maize and johnsongrass. Aust. J. Agric. Res. 19:767-73.

Teetes, G. L. 1985. Insect resistant sorghums in pest management. Insect Sci. Applic. 6:443-451.

USDA. 1874. Guinea grass, pp. 237-40. In: Report of the Commissioner of Agriculture for the year 1873. U.S. Gov. Print. Office, Washington. 496 pp.

USDA. 1875. Johnsongrass, p. 158. In: Report of the Commissioner of Agriculture for the year 1874. U.S. Gov. Print. Office, Washington. 463 pp.

USDA. 1970. Sorghum halepense (L.) Pers. Johnsongrass, pp. 86-7. In: Selected weeds of the United States. USDA Agric. Res. Serv. Agric. Handb. 366. 463 pp.

Vasey, G. 1882. Report of the botanist, pp. 231-55. In: Report of the Commissioner of Agriculture for the years 1881 and 1882. U.S. Gov. Print. Office, Washington. 704 pp.

Vinall, H. N. 1926. Johnsongrass: its production for hay and pasture. U.S. Dept. Agric. Farm. Bull. 1476. 16 pp.
Walter, E. V. 1941. The biology and control of sorghum midge. U.S. Dept. Agric. Farm. Bull. 778. 26 pp.
Wuensche, A. L., G. L. Teetes and J. W. Johnson. 1981. Field evaluation of converted exotic sorghums for resistance to sorghum midge, <u>Contarinia</u> <u>sorghicola</u>. Tex. Agric. Exp. Stn. MP-1484. 30 pp.
Young, W. R. and G. L. Teetes. 1977. Sorghum entomology. Ann. Rev. Entomol. 22:193-218.

7

Sorghum-Corn-Johnsongrass and Banks Grass Mite: A Model for Biological Control in Field Crops

F. E. Gilstrap

Sorghum, wheat and corn are common field crops in Texas, are often planted in adjacent fields, and are part of an ecosystem that typically includes weedy plants. Important arthropod pests in this ecosystem include the Banks grass mite, <u>Oligonychus pratensis</u> Banks (Acari: Tetranychidae); the greenbug aphid, <u>Schizaphis graminum</u> (Rondani) (Homoptera: Aphididae); and sorghum midge, <u>Contarinia sorghicola</u> (Coquillett) (Diptera: Cecidomyiidae). These pests are typically controlled with pesticides, a practice which could result in widespread pest resistance to selected compounds. Resistance is already documented for populations of Banks grass mite (Ward and Tan 1977) and greenbug (Teetes et al. 1975). Biological control of at least one of the named pests will significantly reduce the pesticide load in the local environment, crop production costs and prospects for resistance. However, the temporary nature of these and other annual crops prompts concern among some that annual crops are inadequate for effective biological control. This paper seeks to use the Banks grass mite and the sorghum-corn-wheat-weedy plant ecosystem to illustrate that annual crops are part of a suitably stable environment for development of biological controls.

WHAT IS BIOLOGICAL CONTROL

Biological control simplistically stated is a process wherein pests are killed by their natural enemies and these pests' deaths contribute to production of new enemies. DeBach (1964, p 8) defined biological control as the "study, importation, augmentation and conservation of beneficial organisms for the regulation of population densities of other organisms." Though some (DeBach 1964, 1971; Stern et al. 1959) argue biological control does not require that man is knowingly involved, it is my view that biological control occurs _only_ with purposeful and active participation by man and cannot be fortuitous or accidental. The DeBach (1964) definition for biological control states that the biological control action is _for_ pest regulation (or suppression). Pest control is an integral though implied _purpose_ of the action. Thus, I define biological control as "the result of man acting to employ a biological control tactic (i.e., importation, conservation, or augmentation) which manipulates natural enemies to suppress pest abundance to levels causing no significant damage to man." As argued by Franz (1962), control by natural enemies of which man is unaware or not helping is more appropriately included within the concept and process of "natural control". Each biological control tactic requires that man initiate an action, and that action often requires some understanding by man for optimizing control efficacy.

BIOLOGICAL CONTROL IN EPHEMERAL CROPS

According to Hagen et al. (1976), two attributes of annual crops are: 1) "The ephemeral grain crop is not the ideal ecosystem in which to attempt classical biological control, and the vastness of these annual grain monocultures nearly precludes the use of periodic releases of natural enemies," and 2) "Their (cereals) short cropping period and the rotation practices commonly used in association with their culture create an unstable environment preventing (biological control by) substantial conservation and build-up of natural enemy populations." These two statements do not accurately summarize the prospects for biological control in annual crops, but probably reflect an attitude of many entomologists.

Examples of biological control in ephemeral field crops are relatively few, probably because most biological control effort and attention has historically been devoted to perennial crops. An outgrowth of focus on perennial crops is that perennial crops essentially form the bases for theory and conduct of new biological control programs.

The long history of successful arthropod biological control is heavily weighted to perennial crop ecosystems where established natural enemies are associated with their host (pest) on the same host plant year-around. Such is generally not possible for grain, forage, or range crops. In fact field crops are not the ideal ecosystem for biological control, but neither are these crops the total ecosystem for maintaining the pest and its enemies. Annual crops are actually components of larger agroecosystems which include naturally occurring plants (= weeds), and these agroecosystems have both stability and continuity. The pests and their enemies maintain themselves by occupying other plants when the crop is not available. Whether the crop is a typically large acreage field crop such as grain sorghum, or a smaller acreage high-value and labor intensive crop such as strawberries, the prospects for biological control in annual crops are not as biologically limited as generally thought. However, successful biological control in large acreage field crops requires recognizing and studying the real ecosystem of the pest and its enemies before selecting a biological control tactic and attempting its implementation.

Arthropod mobility is a key consideration for biological control in annual crops. The pests and their natural enemies typically must invade each field each year, either from other crops or from weedy refuges. In either case, the probability is greater for an individual pest to successfully colonize a new location than it is for its natural enemies. Plants are food for the colonizing pest, are numerous and often occur in uniform patterns in relatively large patches (= crop fields). Many of these plants are suitable for pest feeding and reproduction. However, food (= pests) for natural enemies in the early season is nearly always scarce and widely scattered among crop plants. The pest begins reproduction nearly anywhere it attempts colonizing; whereas, the natural enemy must first find the pest to begin feeding and/or reproduction. Usually, the spatial distributions of a pest and its enemies are essentially congruent when the full suppressive impact of an enemy population is realized and of optimal value to crop protection. However, this typically occurs in most annual crops well after the pest successfully colonizes, and often too late to benefit crop protection. At season's end the crops are harvested and residues destroyed, and the newly acquired spatial congruity of the pest and its enemies is also destroyed. Thus, enemies usually must

again invade the crop the following year and, because of
initial spatial discontinuity with the pest and the relatively brief growing season, do not provide adequate control of pests.

REQUISITES FOR BIOLOGICAL CONTROL

The ephemeral nature of annual field crops requires
rigorous adherence to a sequence of steps leading to
implementation of biological control (Table 1). Requisite
#1 deals with obtaining knowledge of the natural enemy
diversity attacking the object pest and is a common
ingredient to all three types of biological control.
Regular samples of all enemies and pests, especially immature stages of pests, are collected from study plots in
order to survey enemy species diversity and to ascertain
their temporal phenology. It is particularly important
that these study plots be continued without interruption
for several cropping seasons, that crop residues be left
intact for at least part of the studies, and that these
plots include "weedy" plants which can naturally host the
pest. The plots should be sampled well after the normal
time for crop harvest, ideally until after the cropping
cycle is initiated the following season. This plot continuity is needed because the researcher needs to know the
impact of not disrupting the pest's enemy(ies) by crop
destruction. Research must initially focus on the crop
pest and its natural enemies in an artificially "natural"
(= undisturbed) crop setting.
Data on Requisite #2, reviewing agronomic or other
crop production practices which may adversely affect
extant enemies, can be collected simultaneous with data
for Requisite #1. Requisite #2 information anticipates
studies manipulating crop production practices to avoid
adversity to enemies and to enhance enemy impact. Requisites 3-6 consist of determining which biological control
action tactic is most promising and amenable to existing
crop production practices. The selected tactic must be
efficacious and acceptable to the farmer, and the ideal is
to modify existing farming practices only if necessary and
only as needed to optimize the impact of the biological
control agent(s).
In the case of importing new natural enemies (Requisite #3), identification of the pest and its enemies (Requisite #1) provides access to taxonomic, geographic

===
Table 1. Requisites for conduct of biological control.

| | Biological control tactic [1] | | |
Biological Control Requisite	Importation	Conservation	Augmentation
ECOSYSTEM DESCRIPTION			
1. Identify extant enemies attacking object pest	+	+	+
2. Review crop practices that may limit enemy efficacy	+	+	+
BIOLOGICAL CONTROL TACTIC APPLICATION			
3. Enemies of same/related pest occur elsewhere	+	-	-
4. Modified production practice allows control by enemies	-	+	-
5. Excess enemies can control pest	-	-	+
6. Determine if enemy production is cost effective	-	-	+

[1] + = studies for this requisite are complete and results were positive; - = studies or information complete with negative results.

distribution and biological literature on the pest and its enemies, or the near relatives of each. If the pest or a near relative occurs in other parts of the world, additional natural enemies can probably be imported from these areas.

Biological control by conservation (Requisite #4) consists of optimizing enemy survival by providing habitat continuity, minimizing insecticide use and/or using only insecticides least toxic to natural enemies, providing or maintaining alternate host plants or host insects for enemy survival, supplementing natural supplies of food for predators or adult parasites, or providing shelters or nesting sites for enemies. However, conservational biological control requires proof that a given enemy can be efficacious when man does not interfere with its natural dynamics. There is little or no economic gain from conserving enemies which cannot contribute significantly to control of a pest or potential pest. Studies for Requisite #4 will suggest adjustments in the timing or extent of crop residue destruction which can significantly conserve important natural enemies. These studies should examine alternate host plants, especially wild plants, which can contribute significantly to habitat continuity needed for survival of natural enemies. If insecticides are commonly applied during crop production, Requisite #4 studies should also determine the effects of these on potentially important natural enemies. Simple screening studies can identify insecticides and application rates least toxic to key enemies operating in the crop.

Developing biological control by augmentation (Requisite #5) entails procedures which are generally more expensive and labor intensive than importation or conservation biological control. Enemies are artificially increased when naturally occurring enemy populations are not great enough when needed for control, or because the enemy occasionally becomes extinct. Additional natural enemies for field release can often be produced in laboratory facilities or in outdoor nurseries. A key to augmentation biological control is cost-effective production of one or more enemy species, sometimes in very large numbers. However, it does little good to release an inherently non-efficacious enemy either very early in the season or in unnaturally large numbers. Thus, an extremely important (and often ignored) aspect of augmentation studies is determining that a numerically augmented enemy can be usefully effective when not limited by reproductive capacity. Furthermore, Requisite #5 studies must establish the numbers of enemies released per unit area and an effective methodology for distributing these enemies.

Finally, costs (if any) of conserving or augmenting natural enemies are generally continuing, and thus must be measured (Requisite #6). These costs should be more than offset by consequent savings to the costs of production.

Generally, the greater the monetary value or input to a crop, the more likely biological control by augmentation or conservation can be cost effective. Importation is theoretically the most cost-effective tactic for implementing biological controls in annual field crops because of its one-time cost.

A CASE STUDY OF BIOLOGICAL CONTROL OF BANKS GRASS MITE

We have worked a number of years towards biological control of Banks grass mite (BGM) on Texas sorghum and corn, and on a common weedy plant, johnsongrass (<u>Sorghum halepense</u> L.). BGM is appropriate for this discussion because its host plant range includes three crop plants typically produced in large acreage fields, and a common weed. A brief description of our work on BGM illustrates the progression of events for Requisites 1-6, and demonstrates the evolution of a biological control program for annual crops. Fig. 1 illustrates what we know about the

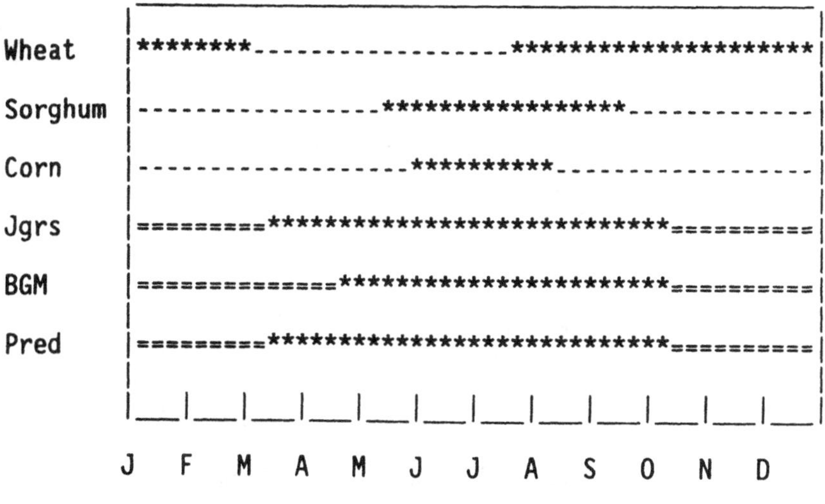

Fig. 1. The ecosystem and phenology of the Banks grass mite (BGM), <u>Oligonychus</u> <u>pratensis</u>, its hosts plants and a phytoseiid (PRED), <u>Amblyseius</u> <u>scyphus</u>, in West Texas.

148

<u>real</u> ecosystem for BGM. Johnsongrass (Jgrs) is present
throughout the year and is rarely disturbed; whereas, the
crop hosts occupied by BGM are completely destroyed after
harvest. Though BGM and its predators (Pred) are able to
sustain themselves beyond the limits of the cropping
season, johnsongrass is the only habitat providing temporal continuity. The BGM and its predators occupy undisturbed stands of johnsongrass year after year, stands
which are overwintering habitat for BGM and its predators.
Our first task was to discover the dynamics of BGM and its
enemies on johnsongrass and on the named crop plants.
Fig. 2 contains data we obtained from johnsongrass plots.

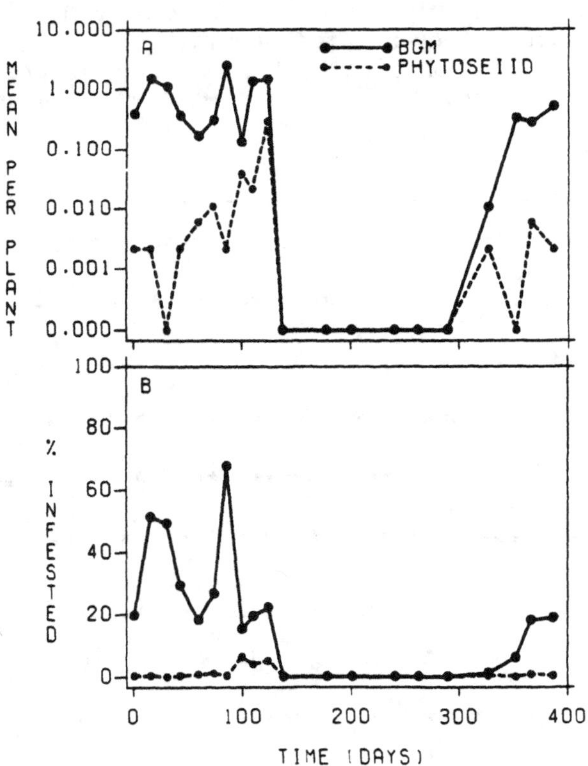

Fig. 2. The population dynamics of the Banks grass mite
(BGM) <u>Oligonychus</u> <u>pratensis</u>, and its phytoseiid predator
<u>Amblyseius</u> <u>scyphus</u>, on johnsongrass in West Texas.

BGM populations on johnsongrass were extremely low, but sufficient to maintain a phytoseiid predator. We learned in our johnsongrass studies that the phytoseiid, <u>Amblyseius</u> <u>scyphus</u> Shuster & Pritchard (Acari: Phytoseiidae), is active and present in most stands of johnsongrass during most months of the year (Gilstrap et al. 1979). Each year BGM and its enemies opportunely leave johnsongrass to occupy new plantings of the three crop plants. These early studies permitted consideration of Requisites #1, #2 and #3. BGM is an arthropod native to North America and is attacked by numerous indigenous enemies. Furthermore, the substantial volume of literature on enemies of spider mites showed that several species of phytoseiid predators are useful for biological control. However, none exhibit characteristics clearly adapting them to overwinter in West Texas. Thus, we initially discounted prospects for classical biological control (Requisite #3) and turned our attention to conservational and augmentation biological control, i.e. Requisites #2, #4 and #5.

We chose to consider conserving enemies of BGM in sorghum because corn and sorghum overlap in availability to BGM, and because sorghum is in the field longest of the two crops. BGM invades sorghum very early in the season and occupies 100% of the plants before day 75, whereas the phytoseiid occupies only about 50% of the sorghum plants on about day 90 (Fig. 3). Successfully invading sorghum is ecologically simplified for BGM as its food (e.g. sorghum) is regularly spaced, planted in relatively large fields, and is generally well-cared for. However, the phytoseiid predator dispersing into sorghum does not possess these advantages. The predator's food source (BGM) in the spring is generally widely scattered among the sorghum plants. BGM begins feeding and reproducing nearly anywhere it lands in the field; whereas, the predators must first find BGM to sustain themselves before beginning reproduction. As the season progresses, more and more predators invade the field and those that succeed in finding BGM also survive to feed and reproduce. A relatively uniform distribution of predators eventually occurs, in part because a few predators successfully establish in the field early in the season and in part because they and their progeny move to new food sources (BGM) as prey become more uniformly distributed. These 2 distribution processes occur simultaneously, improving the odds of encounter between predator and BGM.

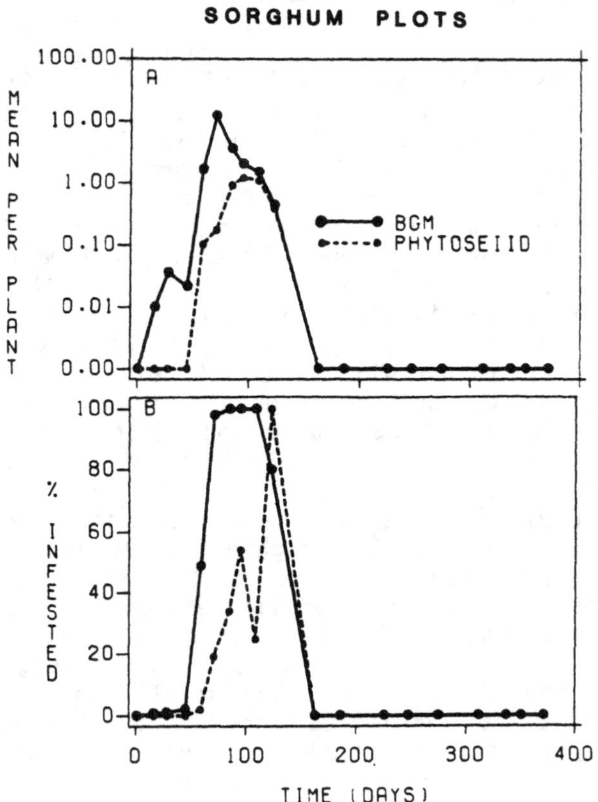

Fig. 3. The population dynamics of the Banks grass mite (BGM) <u>Oligonychus</u> <u>pratensis</u>, and its phytoseiid predator <u>Amblyseius</u> <u>scyphus</u>, on West Texas sorghum when post-harvest crop residues are destroyed.

Grain sorghum in West Texas is typically harvested in the late summer or early fall (at about 120 days post-planting), the residues are destroyed, and in the process BGM and its predators are eliminated from the field. The prey/predator invasion must then repeat itself next year, hopefully with the predators moving into the field early enough to prevent an outbreak of BGM. We harvested our plots on day 110, and then left the post-harvest sorghum plants in the field for overwintering habitat. Because we could find no predators on the leaves, we began dissecting the plants and discovered that phytoseiids were numerous under the leaf sheaths and also inside holes created in stalks by the sugarcane rootstalk weevil, <u>Anacentrinus deplanatus</u> Casey (Coleoptera: Curculionidae). The leaf sheaths and bored areas provided an overwintering habitat for the phytoseiids. Fig. 4 shows the data we collected from our plots where crop residues were not destroyed. When we began dissecting the plants, we discovered an average of more than 20 predators/plant overwintering on

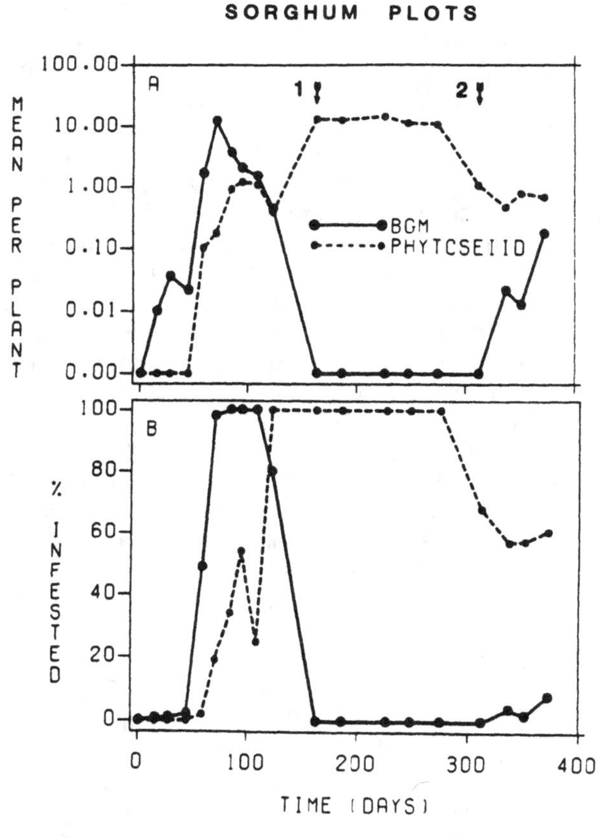

Fig. 4. The population dynamics of the Banks grass mite (BGM) <u>Oligonychus pratensis</u>, and its phytoseiid predator <u>Amblyseius scyphus</u>, on West Texas sorghum when post-harvest crop residues are not destroyed.

these stalks. The phytoseiids on the dead sorghum stalks eventually moved naturally from the stalks of the previous season to the volunteer wheat and johnsongrass of the following season. If a sorghum crop were available the following spring, it is almost certain that the predators would transfer as easily to the new sorghum. Clearly, the phytoseiids can overwinter on standing crop residues when available.

Assuming that _A_. _scyphus_ can be an efficacious control for the BGM, maintaining the previous year numbers and distribution of _A_. _scyphus_ would allow the predators to begin exerting a suppressive influence earlier in the season before BGM becomes substantially more numerous. By providing an overwintering habitat (= the sorghum stubble), we duplicated the essence of the interaction between BGM and _A_. _scyphus_ on johnsongrass. Habitat continuity in the sorghum field conserved the predator's numbers and maintained their distribution across the field for the next year's BGM population. In the context of Requisites listed in Table 1, several cultural modifications could provide the essence of habitat continuity in commercial sorghum including leaving sorghum stalks at periodic intervals across a field, leaving outside rows of sorghum stalks standing through the winter, collecting sorghum stalks after harvest and distributing them across the next season's sorghum seedlings, and/or providing a non-crop overwintering host plant. The objective in all cases would be to provide habitat continuity and take advantage of a useful predator's prior season numbers and distribution, thus promoting control of BGM by naturally occurring predators. We have yet to conduct efficacy studies on _A_. _scyphus_, or to demonstrate any of the listed phytoseiid conservation techniques. These studies are essential to completion of our conservation biological control studies.

Turning to studies on biological control by augmentation (Requisite # 5), corn in West Texas typically develops greater populations of BGM than sorghum. We decided to study predator augmentation on corn because of its reliable spider mite populations. These augmentation studies used two exotic phytoseiid predators produced by a commercial insectary and which were previously demonstrated as having potential for controlling spider mites (Friese and Gilstrap 1982, 1985; Gilstrap and Friese 1985; Oatman et al. 1976, 1977a,b). The predators were <u>Amblyseius</u> <u>californicus</u> (McGregor) and <u>Phytoseiulus</u>

<u>persimilis</u> Athias-Henriot. Our objective was to ascertain predator efficacy without requiring predator overwintering. Fig. 5 is a summary of results from our first attempts at phytoseiid releases on commercial corn (Pickett and Gilstrap 1986a). The predators were released in 1982 at a rate of 1 predator per plant when spider mites were present at a density of about 5/plant. The converging of lines for data at respective study sites shows that numbers of BGM in the release plots (= dashed lines) were similar to those in the control or non-release plots (= solid lines). Only the numbers of BGM in release and non-release plots at Bovina were statistically different from each other. Essentially, the predators were not efficacious in 1982, apparently because the releases were too late. In our 1983 studies (Fig. 6), we released 1 predator/plant when the BGM reached a density of 1 spider mite/plant instead of 5/plant as in the previous year. We obtained striking control in all 3 study fields, with highly significant statistical and plant damage differences between release (= dashed lines) and non-release plots (= solid lines) in each study site. Results of these 2 years of studies showed that both species of phytoseiids can be efficacious in controlling BGM when released at the proper time and at the proper predator/prey ratio. Finally, in 1985 we tested a method for aerial releases of these exotic phytoseiid predators (Pickett et al. 1986). These studies employed a light, fixed wing aircraft fitted with equipment originally designed to release <u>Trichogramma</u> spp. The release methodology distributed insectary produced <u>P</u>. <u>persimilis</u> in uniform to random patterns on commercial corn fields.

The major research tasks remaining are 1) to determine whether indigenous phytoseiids can or cannot control the spider mites in a conservational program, and 2) to ascertain timing and release rates for augmentative releases of exotic phytoseiids. In both cases, a key consideration is to minimize the costs for the biological control activities and yet insure effective spider mite control. Ultimately, the biological control measure(s) demonstrated as efficacious will also need to be proven cost-effective (Requisite #6). The outcome of this latter exercise will depend on the economics of the crops, the selected biological control tactic, and the available alternative control tactics. When our work on BGM is complete, we will have an example to support theory for biological control in annual crops.

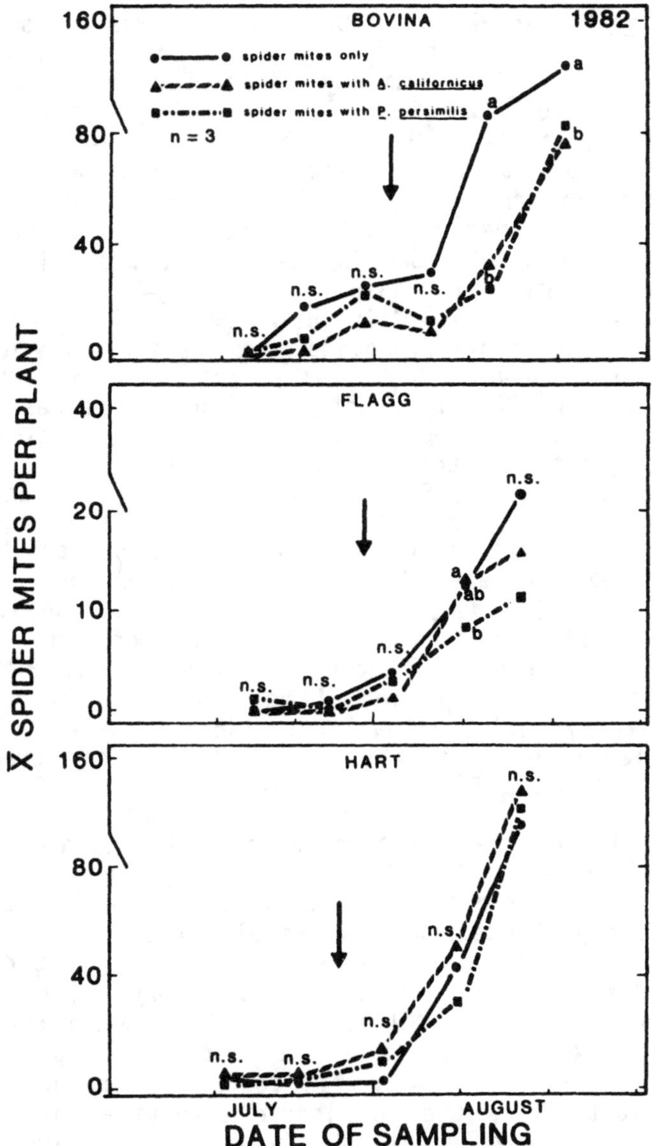

Fig. 5. Densities of spider mites in non-release and predator release plots in 1982. The arrow indicates the date for release of phytoseiids; n.s. indicates none of the means for a given sample date are significantly different. Means followed by different letters (vertically) are significantly different ($P \leq 0.05$).

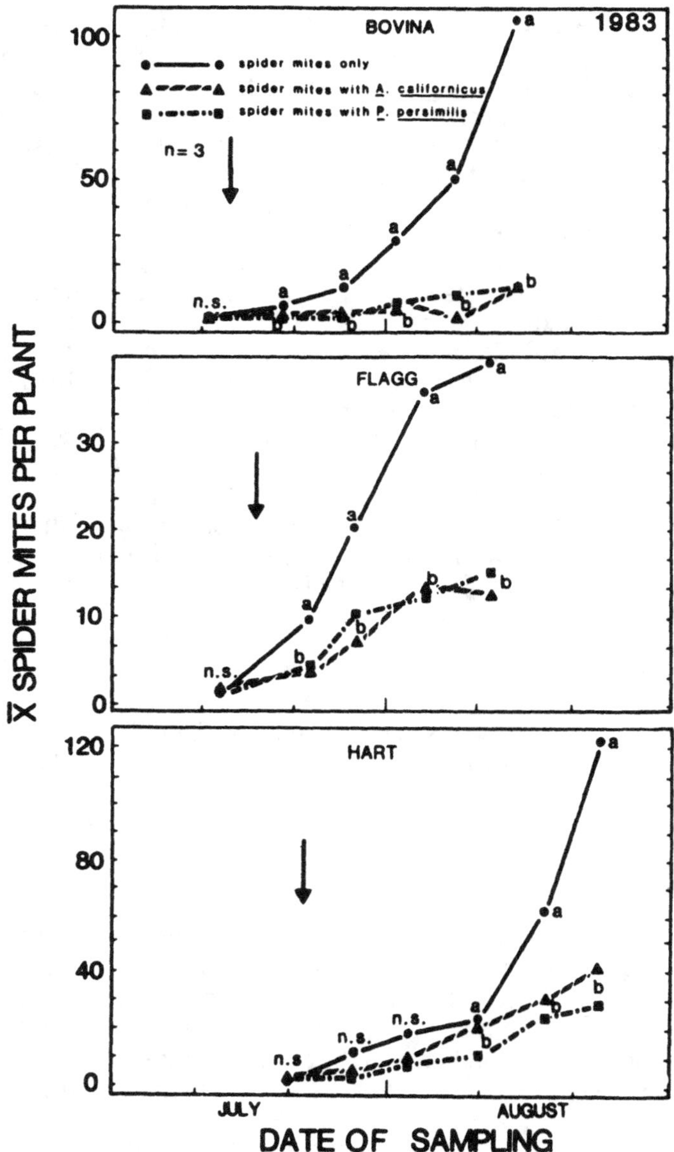

Fig. 6. Densities of spider mites in non-release and predator release plots in 1983. The arrow indicates the date for release of phytoseiids; n.s. indicates none of the means for a given sample date are significantly different. Means followed by different letters (vertically) are significantly different ($P \leq 0.05$).

From progress in our work, I remain convinced that annual field crops are amenable to cost effective biological control. To recognize and address the true perennial nature of the annual crop agroecosystem significantly improves prospects for biological control in these crops. Clearly, our work on biological control of spider mites is far from complete. When these studies are complete, we will have evaluated the efficacy of the dominant indigenous predators and ways to conserve them, and will also have evaluated and developed methodology for augmentative releases of exotic predators. All of these studies are needed to establish a scientific basis for an IPM specialist to incorporate biological control tactics into a crop protection strategy. Our progress to date on biological control of BGM (Pickett and Gilstrap 1984, 1986b), greenbug (Gilstrap et al. 1984; Kring and Gilstrap 1982, 1983, 1984, 1986; Kring et al. 1985; Summy and Gilstrap 1983; Summy et al. 1979) and sorghum midge (Brooks and Gilstrap 1985, 1986) gives cause for optimism regarding prospects for biological control in annual crops.

ACKNOWLEDGMENT

This research was conducted by the Tex Agric. Exp. Stn., with partial support from a U.S. AID grant AID/DSAN/XII/G-0149.

REFERENCES CITED

Brooks, G. W., and F. E. Gilstrap. 1985. Parasitism of sorghum midge in johnsongrass in major vegetational regions of Texas. Southw. Entomol. 10: 171-5.

Brooks, G. W., and F. E. Gilstrap. 1986. Seasonal incidence and abundance of sorghum midge and associated parasites in johnsongrass. Southw. Entomol. 11: 119-24.

DeBach, P. 1964. The scope of biological control. Chapter 1 in: Biological Control of Insect Pests and Weeds. Reinhold Publ. Corp., New York. 844 pp.

DeBach, P. 1971. Fortuitous biological control from ecesis of natural enemies. IN: Entomological Essays to Commemorate the Retirement of Professor K. Yasumatsu, pp. 293-307. Hokuryukan Publ. Co. Ltd. Tokyo. 389 pp.

Franz, J. M. 1962. Definitions in biological control. XI. Intl. Kong. Entomol. Wien (1960). Verhandl Band II., Sect. XIII, 670-4.

Friese, D. D., and Gilstrap, F. E. 1982. Influence of prey availability on reproduction and prey consumption of _Phytoseiulus persimilis_, _Amblyseius californicus_, and _Metaseiulus occidentalis_ (Acarina: Phytoseiidae). Internat. J. Acarol. 8: 85-90.

Friese, D. D., and Gilstrap, F. E. 1985. Prey requirements and developmental times of three phytoseiid species predaceous on spider mites. Southw. Entomol. 10: 83-88.

Gilstrap, F. E., and D. D. Friese. 1985. The predatory potential of _Phytoseiulus persimilis_, _Amblyseius californicus_, and _occidentalis_ (Acarina: Phytoseiidae). Internat. J. Acarol. 11: 163-8.

Gilstrap, F. E., T. J. Kring, and G. W. Brooks. 1984. Parasitism of aphids attacking aphids associated with Texas sorghum. Environ. Entomol. 13: 1613-17.

Gilstrap, F. E., K. R. Summy, and D. D. Friese. 1979. The temporal phenology of _Amblyseius scyphus_, a natural predator of Banks grass mite in West Texas. Southw. Entomol. 4: 27-34.

Kring, T. J., and F. E. Gilstrap. 1982. Within field distribution of greenbug and its parasitoids in winter wheat. J. Econ. Entomol. 76: 57-62.

Kring, T. J., and F. E. Gilstrap. 1983. Population dynamics of the greenbug and its parasitoids on winter wheat in Central Texas. Tex. Agr. Exp. Sta. Prog. Rept. PR-4140. 10 pp.

Kring, T. J., and F. E. Gilstrap. 1984. Efficacy of the natural enemies of grain sorghum aphids (Homoptera: Aphididae). J. Kansas Entomol. Soc. 57: 460-7.

Kring, T. J., and F. E. Gilstrap. 1986. Beneficial role of corn leaf aphid, _Rhopalosiphum maidis_ (Fitch), in maintaining _Hippodamia_ spp. (Coleoptera: Coccinellidae) in grain sorghum. J. Crop Protection 5: 125-8.

Kring, T. J., F. E. Gilstrap, and G. J. Michels, Jr. 1985. Role of indigenous coccinellids in regulating greenbug, _Schizaphis graminum_ (Rondani) on Texas grain sorghum. J. Econ. Entomol. 78: 269-73.

Oatman, E. R., F. E. Gilstrap, and V. Voth. 1976. Effect of different release rates of <u>Phytoseiulus persimilis</u> (Acarina: Phytoseiidae) on the two spotted spider mite on strawberry in southern California. Entomophaga 21: 269-73.

Oatman, E. R., J. A. McMurtry, F. E. Gilstrap, and V. Voth. 1977a. Effect of releases of <u>Amblyseius californicus</u>, <u>Phytoseiulus persimilis</u>, and <u>Typhlodromus occidentalis</u> on the two spotted spider mite on strawberry in southern California. J. Econ. Entomol. 70: 45-7.

Oatman, E. R., J. A. McMurtry, F. E. Gilstrap, and V. Voth. 1977b. Effect of releases of <u>Amblyseius californicus</u> on the two spotted spider mite on strawberry in southern California. J. Econ. Entomol. 70: 638-40.

Pickett, C. H., and F. E. Gilstrap. 1984. Phytoseiidae associated with Banks grass mite infestations in Texas. Southw. Entomol. 9: 125-33.

Pickett, C. H., and F. E. Gilstrap. 1986a. Inoculative releases of phytoseiids (Acari) for the biological control of spider mites (Acari: Tetranychidae) in corn. Environ. Entomol. 15: 790-4.

Pickett, C. H., and F. E. Gilstrap. 1986b. Natural enemies associated with spider mites (Acarina: Tetranychidae) infesting corn in the High Plains of Texas. J. Kansas Entomol. Soc. 59: 524-36.

Pickett, C. H., F. E. Gilstrap, R. K. Morrison, and L. F. Bouse. 1986. Release of predatory mites (Acari: Phytoseiidae) by aircraft for the biological control of spider mites infesting corn. J. Econ. Entomol. 80: 906-10. ..   # 3065

Stern, V. M., R. F. Smith, R. van den Bosch, and K. S. Hagen. 1959. The integration of chemical and biological control of the spotted alfalfa aphid. The integrated control concept. Hilgardia 29(2): 81-101.

Summy, K. R., and F. E. Gilstrap. 1983. Facultative production of alates by greenbug and corn leaf aphid, and implications in aphid population dynamics. J. Kansas Entomol. Soc. 56: 434-40.

Summy, K. R., F. E. Gilstrap, and S. M. Corcoran. 1979. Parasitization of greenbug and corn leaf aphid in West Texas. Southw. Entomol. 4: 176-80.

Teetes, G. L., C. A. Schaefer, J. R. Gipson, R. C. McIntyre, E. E. Latham. 1975. Greenbug resistance to organophosphorus insecticides on the Texas High Plains. J. Econ. Entomol. 68: 214-6.
Ward, C. R., and F. M. Tan. 1977. Organophosphate resistance in the Banks grass mite. J. Econ. Entomol. 70: 250-2.

8

Rice Insect Pests
and Agricultural Change

M. A. Loevinsohn, J. A. Litsinger, and E. A. Heinrichs

TRADITIONAL RICE CULTURE

Man first domesticated rice, <u>Oryza</u> <u>sativa</u> L., some 10,000 years ago in river valleys in South and Southeast Asia (Chang, 1976). Though upland varieties adapted to well-drained conditions were selected with time, rice cultivation has remained for the most part tied to environments where standing water can be maintained. Where such conditions did not exist, extravagant effort was sometimes expended to create them, as in the rice terraces of Bali, Nepal, and the northern Philippines. Its dependence on water has enabled rice to escape many weeds intolerant of anaerobic conditions, but has also meant that few crops can follow wet rice in a rotation because of the impervious hardpan and radically disturbed soil structure that typically arises (De Datta, 1981). Rice is, as perhaps no other, a crop destined for monoculture.

The traditional indica varieties that were grown in most of tropical Asia before the 1960's were photoperiod-sensitive, generally flowering during the short days at the end of the monsoon rains from October to December (Vergara and Chang, 1976). These varieties were often well adapted to low or moderate soil fertility and responded to nitrogenous fertilizer with excessive vegetative growth, leading to lodging. They were for the most part leafy in appearance, with a high leaf area index (Yoshida, 1981). Though leading to a poor photosynthetic efficiency, these characteristics likely conferred a degree of tolerance to defoliating insects, which were often the dominant pests.

The rice crop was for the most part established by transplanting into puddled soil, though direct seeding has traditionally been practiced in areas of unpredictable rainfall and in upland cultivation. The amount of care expended on the crop after transplanting varied enormously and appears to have been related to the man-land ratio. In parts of Java and Bali for example, high population densities were reflected in close attention to water control and in careful weeding (Geertz, 1963). In the Cagayan Valley of the Philippines, in contrast, the crop was practically abandoned until harvest. Insect control was similarly variable in its intensity. Traditional practices included a variety of cultural techniques, removal of insects by hand or nets, pouring kerosene on the flood water, and the use in different ways of indigenous plants with insecticidal properties (Litsinger et al., 1980). By the mid-1960's, farmers in many areas had first-hand experience in the use of synthetic insecticides (Kenmore et al., 1987) but these were in general little used due to the low yields, scarcity of credit, and sporadic pest pressure.

MODERN RICE CULTURE

Modern Varieties

Rice is the major food of one-third of the world's population and the principal cereal in Asia. The low rice yields of tropical Asia in the 1950's set against rapidly increasing human populations prompted both national governments and international aid agencies to reflect on how rice production might be rapidly increased (Chandler, 1982). The success of Japan in increasing production through the breeding of japonica rices responsive to increased fertilization provided a model for indica rice improvement.

At the International Rice Research Institute (IRRI), the tall vigorous traditional variety Peta from Indonesia was crossed with the high-tillering, semi-dwarf Dee-geo-woo-gen from China to produce IR8. The new plant type captured most of the best traits of both parents to produce a variety with high yield potential. IR8 was short with stiff culms which could support the weight of the grain when fertilized. Its high tillering capability allowed the plant to fill in the empty spaces in the field. IR8 also had moderate seed dormancy and was photoperiod insensitive. As of 1986, 29 IR varieties have been released in the Philippines for lowland irrigated environments (Table 1).

Table 1. Resistance of IR varieties to insect pests.[a] IRRI, 1986.

Variety	Maturity (DAS)	N. lugens biotype			Nephotettix virescens	Sogatella furcifera	Scirpophaga incertulas	Chilo suppressalis	Hydrellia philippina
		1	2	3					
IR5	130	S	S	S	MR	S	S	S	S
IR8	125	S	S	S	MR[d]	S	S	S	S
IR20	121	S	S	S	MR[d]	S	MR	R	S
IR22	119	S	S	S	S	S	S	S	S
IR24	118	S	S	S	MR	S	S	S	S
IR26	121	R	S	R	MR	S	S	MR	S
IR28	104	R	S	R	R	S	S	S	S
IR29	112	R	S	R	R	S	S	S	S
IR30	105	R	S	R	R	S	S	MR	S
IR32	130	R	R	MR[c]	MR	S	S	MR	S
IR34	122	R	S	R	R	S	S	MR	S
IR36	110	R	R	MR[c]	MR	S	MR	R	S
IR38	123	R	R	MR[c]	MR	S	S	MR	S
IR40	130	R	MR	S	MR	S	MR	R	MR
IR42	132	R	R	S	MR	S	S	MR	S
IR43	120	S	S	S	MR	S	S	MR	S
IR44	123	R	R	MR[c]	MR	S	S	R	S
IR45	120	R	S[b]	R	MR[d]	S	S	S	S
IR46	112	R	S[b]	R	MR[d]	S	S	S	S
IR48	127	R	R	S	MR	MR	S	S	S
IR50	107	R	MR	R[c]	R	S	MR	R	S
IR52	117	R	R	MR	R	MR	S	MR	S
IR54	120	R	R	S	R	S	MR	MR	S
IR56	106	R	R	R	R	S	S	MR	S

Table 1. Continued

IR58	108	R	R	R	R	S	S	MR	S
IR60	110	R	R	R	R	MR	S	MR	S
IR62	110	R	R	R	R	MR	S	MR	S
IR64	115	R	MR	R	R	S	S	e	S
IR65	115	R	R	R	R	MR	MR	e	S

[a]Based on replicated experiments. Hopper resistance based on greenhouse evaluation of seedlings; yellow and striped stem borer resistance based on a screenhouse evaluation of 40-70-day-old plants; leaffolder and caseworm resistance based on the reaction of 30 and 11-day-old plants in the greenhouse; whorl maggot resistance based on field observations at 30 days after transplanting. Varieties with ratings as based on the standard evaluation system of 1-3 are considered resistant (R), 5-7 moderately resistant (MR) and 9 susceptible(s).

[b]IR46 has field resistance to biotype 2.

[c]Reaction to biotype varies, occasionally being susceptible and often resistant.

[d]Occasional susceptible reactions. Least resistant of the IR varieties except for IR 22 which is highly susceptible.

[e]Tests still to be conducted.

The newer varieties have yield potentials similar to that of IR8 yet possess broader resistance to stresses and better eating quality.

Insect pests and the diseases they vector have played a prominent role in the proliferation of IR varieties. Outbreaks of brown planthopper <u>Nilaparvata</u> <u>lugens</u> (Stål) (Dyck and Thomas, 1979) and green leafhopper <u>Nephotettix</u> <u>virescens</u> (Distant) (Sogawa, 1976) have occurred in rapid succession after the release of IR8 in 1966. The Philippines represents a case in point.

In 1970-72 outbreaks of green leafhopper and tungro virus covering hundreds of thousands of hectares plagued the Philippines. IR20, with TKM6 parentage (imparting moderate resistance to green leafhopper and moderately susceptible to tungro virus) was released in 1969 and eventually became sufficiently widespread to quell these problems but proved susceptible to brown planthopper. IR26, released in 1973 was quickly overcome by a new biotype of brown planthopper and outbreaks of green leafhopper occurred as well in 1974. IR36, released in 1976, with Ptb 18 parentage (imparting moderate resistance to both green leafhopper and tungro) and the <u>bph</u>2 gene stood up until 1982, at which time biotype 3 of brown planthopper caused small outbreaks in Mindanao and tungro and grassy stunt II epidemics occurred in several areas. This prompted the release of IR56 and subsequent varieties with the <u>bph</u>3 gene for brown planthopper and Gam Pai parentage (imparting resistance to green leafhopper and tungro). In 1986, the <u>bph</u>3 gene was overcome on Java and tungro repeatedly occurred on IR56, IR58, IR60, IR62, and IR64 in the Philippines.

Elite lines with improved resistance to green leafhopper and tungro will be released soon as new varieites by the Philippine Seed Board, but it is clear that varietal development is on a treadmill: each advance in breeding for resistance is soon countered by newly evolved virulence in the insect or pathogen population.

The introduction of germplasm from wild rices (<u>Oryza</u> spp.) is a breeding approach for the short- to medium-term that provides new resources to the breeder in this struggle. Breeding methods have improved in recent years, allowing more wide crosses to be made between <u>Oryza</u> spp. and <u>O</u>. <u>sativa</u>. Resistance to grassy stunt virus I has been introduced into rice from <u>O</u>. <u>nivara</u> Sharma et Shastry and screening of the wild rice collection in the IRRI germplasm bank against the major pests reveals high rates of resistance (Table 2). However, the recent history of

Table 2. Evaluation of rice *Oryza sativa* and wild rice *Oryza* spp. accessions in the IRRI germplasm collection for resistance against insect pests of rice. IRRI, 1962–1986.

Insect	*Oryza sativa*		Wild rices	
	Screened	Resistant	Screened	Resistant
	(No.)	(%)	(No.)	(%)
Brown planthopper *Nilaparvata lugens*				
Biotype 1	48,542	2.5	433	46.0
Biotype 2	15,114	1.9	432	38.0
Biotype 3	16,642	1.7	435	39.0
Green leafhopper *Nephotettix virescens*	53,896	2.5	435	53.0
Whitebacked planthopper *Sogatella furcifera*	52,041	0.9	436	46.0
Yellow stem borer *Scirpophaga incertulas*	24,314	0.1	424	18.0
Striped stem borer *Chilo suppressalis*	15,000	0.2	240	5.4
Whorl maggot *Hydrellia philippina*	22,676	0.01	324	2.2
Leaffolder *Cnaphalocrocis medinalis*	21,750	0.5	432	2.1

introduction and rapid breakdown of resistance indicates the need for new approaches to the use of genetic resources appropriate to small-hold agriculture.

Two tactics that offer some hope and that have attracted recent attention are multilines (Anonymous, 1977; Chin and Husin, 1982) and varietal rotations (Anonymous, 1982; Manwan et al., 1985). The development of multilines will require an earnest breeding effort to assemble agronomically compatible lines and an on-farm testing program to ensure acceptability to farmers. The durability of varietal rotations will hinge on the persistence of virulence genes in insect or pathogen populations after several generations of exposure to resistant varieties. In much of Asia, it may prove impossible to ensure coordinated rotations on the poorest and most isolated farms where seed is seldom purchased.

In addition to resistance to pest attack, an important characteristic of modern varieties is their more rapid maturation. As the populations of most major pests typically increase through the season until the cessation of vegetative growth or senescence limits the infestation, shorter duration varieties planted in synchrony with the prevailing cropping pattern may suffer reduced attack (Fig. 1). Early maturity thus constitutes a simple form of resistance or, more accurately, evasion (Painter, 1951). There is evidence, however, that the faster vegetative growth of very short duration varieties may limit their ability to compensate for pest damage (Litsinger et al., 1987).

Modern Rice Production Practices

A series of changes in rice cultural practices have accompanied the dissemination of modern varieties. Perhaps most importantly, photoperiod insensitivity and reduced growth duration have made double or even triple cropping possible where irrigation or rainfall is adequate. In turn, the availability of these varieties has made investment in large-scale irrigation schemes attractive and many have been constructed within the last 20 years throughout tropical Asia.

The spread of dry season cultivation has had a number of consequences for rice insects. It has led, firstly, to shifts in the composition of the pest fauna. In tropical areas, most pest species are not limited in their occurrence by seasonal factors and appear whenever rice is planted. A notable exception is the Asian white rice borer _Scirpophaga_

168

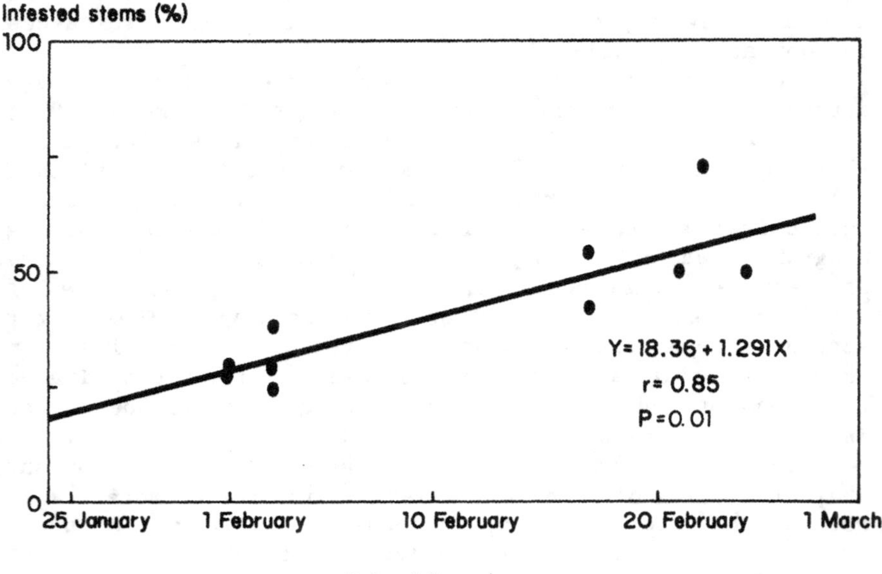

Figure 1. Infestation by the African white rice borer, *Maliarpha separatella*, in relation to the date of flowering of 10 rice varieties. Mahitsy, Madagascar, 1985. (Data courtesy FOFIFA)

innotata (Walker) which has largely disappeared from intensively cultivated areas. Its larvae aestivate in the stubble and are destroyed by dry season land preparation.

Changed physical conditions have been responsible for other shifts. Species dependent on standing water such as whorl maggot *Hydrellia philippina* Ferino, rice caseworm *Nymphula depunctalis* (Guenee), green semilooper *Naranga aenescens* Moore, and hairy caterpillar *Rivula atimeta* (Swinhoe) -- have become more abundant than in traditional rainfed cultivation. Conversely, pests like the mole cricket *Gryllotalpa orientalis* Burmeister, which plagues fields with inadequate water control and those in dryland areas, have diminished in importance.

Concomitantly with the expansion of dry season rice cultivation, there has been a reduction in the area planted to alternate crops such as maize and grain legumes as well as in the cover of wild grasses in fields formerly left fallow in the dry season. This vegetational change has favored rice specialist pests with monophagous and oligophagous habits at the expense of the polyphagous. This can be seen in the reduced importance of generalist leaf feeding pests of grasses such as cutworms, _Spodoptera_ spp., armyworms, _Mythimna_ spp., and the oriental migratory locust, _Locusta migratoria manilensis_ (Meyen) which has seen many of its breeding grounds transformed into rice lands. Similar trends can also be seen within the guild of stem boring insects. In North Krian, Malaysia, the proportion of the area devoted to rice cultivation that was double cropped rose rapidly with the arrival of irrigation and exceeded 50% by the late 1960's. This was reflected in reduced abundance of the polyphagous dark-headed borer, _Chilo polychrysus_ (Meyrick), and an increase in that of the monophagous yellow rice borer, _Scirpophaga incertulas_ (Walker) (Fig. 2). Much the same pattern has been described in Laguna, Philippines, where the yellow rice borer became markedly more abundant than the striped stem borer, _Chilo suppressalis_ (Walker), a

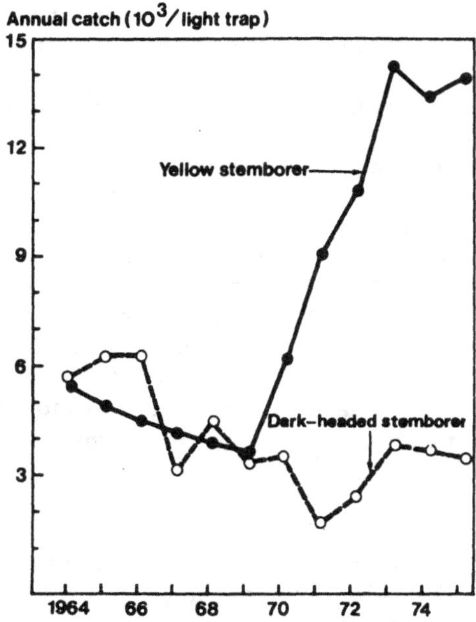

Figure 2. Annual light trap collections showing the replacement of dark-headed stem borer _Chilo polychrysus_ with yellow stem borer _Scirpophaga incertulas_ with the expansion of double rice cropping in Titi Serong, Malaysia, 1964-1975. (Loevinshohn, 1984)

species known to feed on dryland crops such as maize and sugarcane (Loevinsohn, 1984).

The rise to dominance of rice specialist pests has been most clearly seen in the case of the plant sucking Homoptera. Two genera have been responsible for major losses in recent years in almost every nation of tropical Asia: the nearly monophagous brown planthopper, and the green leafhopper complex, in particular <u>Nephotettix virescens</u>, the most monophagous of the tropical species and the most efficient vector of tungro virus. The brown planthopper rose from the status of a minor pest in the mid-60's to that of a major determinant of yield, responsible, together with its associated viruses, for losses estimated at 365,000 metric tons of milled rice in 1976-77 in Indonesia, equivalent to the annual consumption of some 3 million people (Dyck and Thomas, 1979).

In general, the more specialized pest fauna that is now typical of intensively cultivated areas is also one that is responsible for increased crop losses in both absolute and proportional terms. Table 3 provides estimates of losses in environments of differing cropping intensities in the Philippines.

The increased abundance and economic impact of major pest species that has followed the intensification of rice cultivation has constituted a professional and scientific challange to entomologists, one that has not been fully met. Though it has been suggested here that changes in the duration and extent of cultivation have been chiefly responsible for these developments, this has by no means been accepted by all. The dimensions of the disagreement were clearly seen at a 1977 symposium that sought a consensus on the causes of brown planthopper outbreaks and on strategies to deal with them (Anonymous, 1979). Participants pointed to a number of roughly contemporaneous and often related changes in rice cultivation and in the characteristics of the rice plant. These are discussed briefly.

The use of nitrogenous fertilizer increased markedly in Asia following the adoption of nitrogen responsive modern varieties and in response to government-sponsored loan programs. Numerous experiments have shown a positive correlation between the numbers or damage caused by certain insect species and applied nitrogen. Closer plant spacing was made possible by the more erect growth habit of modern varieties and this has been found to create a more favorable habitat for the buildup of the brown planthopper. Insecticide use increased markedly following the

Table 3. Estimated crop losses to insect pests and associated diseases at 5 sites (provinces) in the Philippines. Losses were estimated as the difference in wet season yield between check plots and those treated nine times. (Loevinsohn, 1984).

Site	Mean rice crops/yr	Yield loss	
		Tons/ha	% Protected yield
Batangas	1.0	0.25	7
Pangasinan	1.0	0.55	12
Cagayan	1.0	0.50	14
Iloilo	1.5	0.70	24
Nueva Ecija	1.9	2.37	33

dissemination of modern varieties, again often as a result of government programs, and resurgence of brown planthopper, green leafhopper, and rice leaffolder _Cnaphalocrocis medinalis_ Guenee (Kiritani, 1979; Reissig et al., 1982; Kenmore et al., 1984).

The inability heretofore to critically test these hypotheses has meant that it has been impossible to launch a concerted attack on the causes of increased pest abundance and that control has largely entailed coping with its consequences. Farmers have had no choice but to rely on broad spectrum insecticides, with their attendant rising real costs and risks to health and the environment, and on varietal resistance that has in many cases not proved durable and which in any case is available only for certain pest species.

Cropping Intensity and Synchrony

A recent analysis however may provide some insight into the causes of increased pest densities (Loevinsohn, 1984). Examination of light trap records spanning the period of agricultural change in the Philippines and Malaysia, as well as of contemporary records from areas in the Philippines that differ in the intensity of cultivation, suggests that higher pest densities have generally been associated with lower rates of pest increase per season, as is to be expected from the numerical and functional response of natural enemies. In all cases, however, there has been a trend through time or with intensification toward an increase in the rate of change between wet seasons, that is, toward a reduced rate of decline during the intervening period.

It is important to note that of the factors proposed to account for the increased abundance of major pests, most, including the effects of fertilizer and pesticide, would act through the within-season rate of increase, while only two affect the rate of change between seasons: 1) the impact of double cropping and 2) asynchrony in cultivation between farms. Both reduce the length of the fallow that pests must endure. These latter factors are in many instances related: dependence on irrigation has introduced new sources of variation in the timing of cultivation. Distance from the canal, structural constraints on irrigation capacity, and difficulties with drainage may now oblige farmers to either advance or retard their planting where, under rainfed conditions, they had relied on rainfall that instigated planting over large areas.

Figure 3 shows that for two major pests in Malaysia, increase in the rate of population change between wet seasons has been closely correlated with the spread of irrigation.

The impact of double cropping on the numbers of yellow rice borer and brown planthopper at 10 sites in the Philippines is illustrated in Fig. 4. The fitted equations suggest that for both insect species, the increase has been exponential. (Note that a simple doubling of annual abundance with a shift from a single to double cropping would yield a slope of 0.693 on semi-logarithmic plot. In fact, both slopes are significantly greater than this value).

Other factors generally associated with double cropping, such as greater use of agrochemicals or the planting of modern varieties, were not found to account for significant additional variation in pest abundance. The same applies to differences between sites in the area devoted to rice cultivation, although the range of values is small. In general in Asia, the contribution to increased production resulting from expansion of the area under cultivation has been minimal in recent years (Barker and Herdt, 1985), although the experience elsewhere, notably in South America and Africa, has been different. On theoretical grounds, it has been argued that the response of pest populations to an increase in the extent of cultivation should under most circumstances be close to or less than

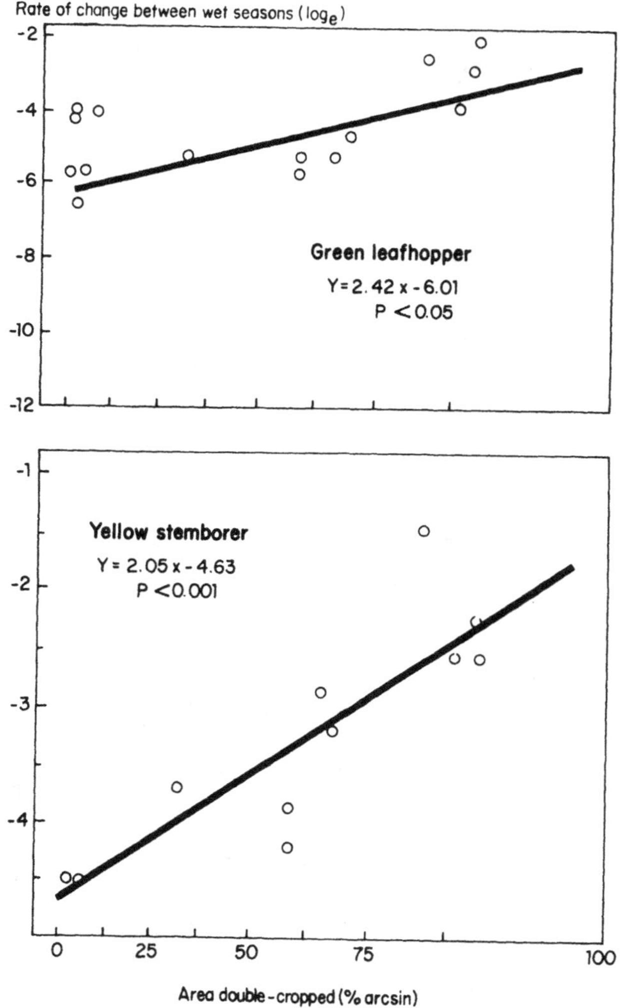

Figure 3. Annual light trap collections of green leafhopper *Nephotettix virescens* and yellow stem borer *Scirpophaga incertulas* spanning the period of change from single to double cropped rice with the expansion of irrigation in Titi Serong, Malaysia, 1965–1975. The rate of change between wet seasons is the ratio of the insect population at the beginning of one wet season to the peak of the preceding wet season. (Loevinshohn, 1984)

174

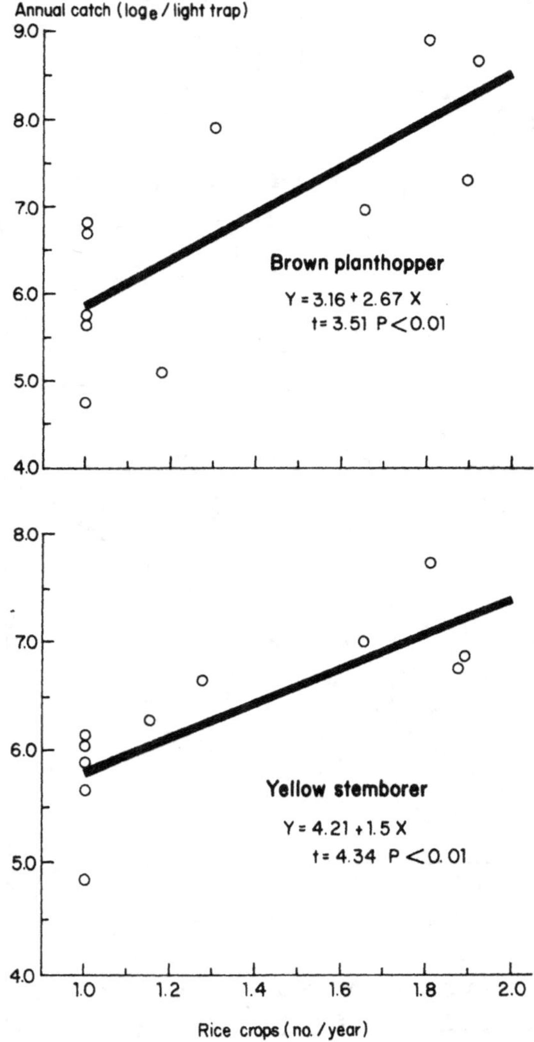

Figure 4. Annual light trap collections of brown planthopper *Nilaparvata lugens* and yellow stem borer *Scirpophaga incertulas* from 10 rainfed and irrigated rice sites varying in the mean number of rice crops per year. Philippines. (Loevinshohn, 1984)

algebraic (Loevinshohn, 1984), that is, doubling of the proportion of area devoted to rice should result in less than a doubling of local pest abundance.

The independent contribution to variation in pest abundance of the asynchrony of cultivation can be estimated in environments where the number of crops grown per year is approximately uniform. In such an area in Nueva Ecija province, Philippines, the logarithm of seasonal total catch in farmer-operated light traps of 8 of 9 pest species was found to be significantly related to the standard deviation of planting date, a measure of asynchrony (Loevinsohn, 1984). Two such relationships are illustrated in Fig. 5. Asynchrony was estimated within areas of from 0.2 to 2.0 km radius around the light traps. The radius within which the best correlation between asynchrony and pest abundance was obtained differed among species but was relatively constant over seasons and appeared to reflect relative dispersal capacities (Fig. 6).

It was also found, as expected, that the rate of response to asynchrony, the change in log seasonal abundance per day of standard deviation of planting date, was correlated with estimates of the species' maximum rate of increase. Theory suggests, and some experimental results indicate, that there should be no increase in the abundance of a pest species if the overall range of variation in planting date is of less than a generation length within a distance that the insect is capable of flying.

Generation length and dispersal range provide the basic temporal and spatial criteria with which to assess the impact of unintentional delay, for example that caused by irrigation scheduling. They also permit the design of schemes for the synchronization of planting dates that would place the least possible strain on farmers' resources for preparing the land and on the labor available for transplanting the crop (Fig. 7). It is significant that in Nueva Ecija, areas well served with irrigation are able to transplant their crops within a period that does not permit the completion of an additional generation of the major pests within a distance equivalent to their median dispersal distances.

RICE ECOSYSTEMS AND PEST MANAGEMENT

We believe, on the basis of the foregoing, that efforts to limit the carrying capacity for pests of rice agroecosystems must underlie integrated control programs. The fact that most rice pests respond positively to the

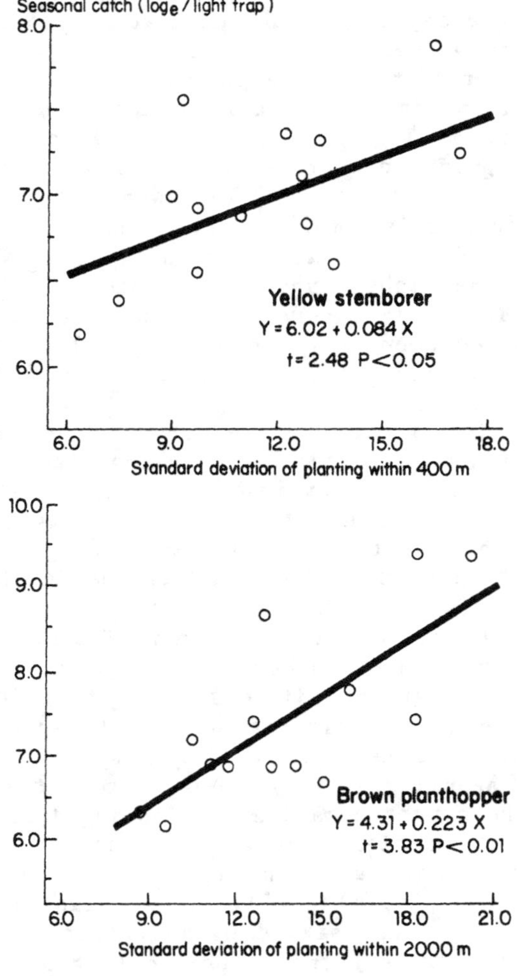

Figure 5. Seasonal totals of yellow stem borer *Scirpophaga incertulas* and brown planthopper *Nilaparvata lugens* collected in kerosene light traps at 14 irrigated rice sites in relation to asynchrony of planting. The relationship is illustrated with the radius from each light trap that provided the best fit between asynchrony and log seasonal numbers, an index of the species effective dispersal range. Nueva Ecija, Philippines, Wet Season, 1981. (Loevinsohn, 1984)

duration of rice availability, both due to multiple cropping and to the asynchrony with which crops are established, suggests that reduction of either would result in diminished pest populations. The feasibility of such a reduction however must be carefully considered. A retrenchment to one crop a year in areas of double cropping would have major impacts on the income of farmers and landless laborers as well as on the availability of food to urban consumers. Its effects cannot be predicted until specific alternatives to double cropping are formulated and tested. Planting an upland crop, such as a legume, in place of the dry season rice crop may be economical in certain areas, and if widely practiced, would impose a prolonged fallow on rice specialist pests, breaking their life cycles.

Such questions of agroecosystem design must be considered from a multidisciplinary perspective, such as

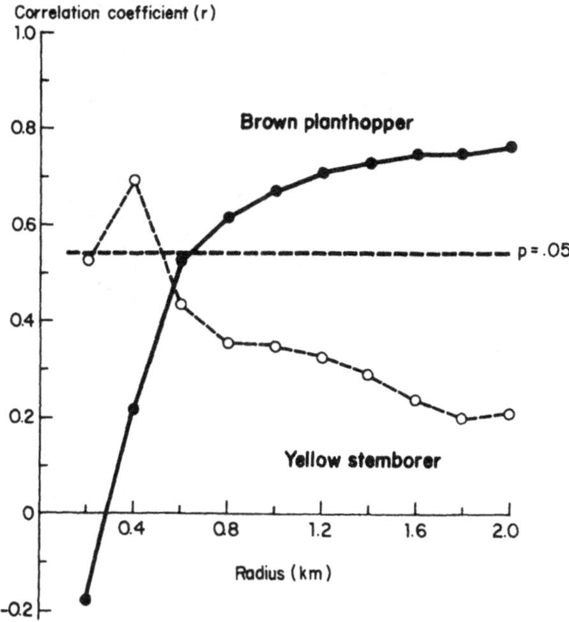

Figure 6. The correlation between seasonal light trap catch of brown planthopper *Nilaparvata lugens* and yellow stem borer *Scirpophaga incertulas* and asynchrony calculated for areas of increasing radii. Nueva Ecija, Philippines, Wet Season, 1981. (Loevinsohn, 1984)

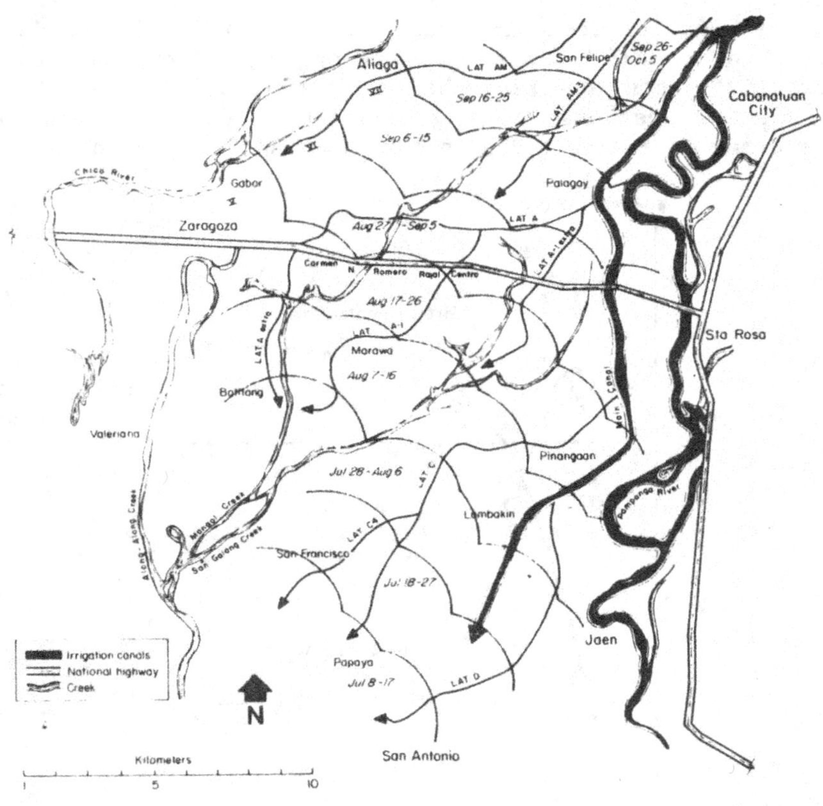

Figure 7. A zonal schedule of planting dates eases the power, labor, water, and credit constraints that synchronization of planting over a wide area would produce. In this example, planting begins at the downstream end of the 17,000 ha upper Pampanga River Integrated Irrigation System and proceeds in 2.5-km-wide zones planted at 10-day intervals.

that offered by farming systems research (Zandstra et al., 1981). However, we suggest that efforts to further intensify rice cultivation to three crops per year where irrigation makes this possible should be resisted. Extrapolation of available information suggests that both pest densities and losses to the crop would increase, and observations in the limited areas where three crops are grown bear this out.

Synchronization of planting appears an inherently less problematic strategy than limitation of cropping intensity. Indeed, the improvements in irrigation delivery and drainage that would make it possible are likely to have independent and positive effects on yield, making the necessary efforts more attractive. Unlike other cultural control methods, such as alteration of planting date, tillage practices, crop density or water depth, the effects of synchronization are not pest specific and may include as well reduction of several diseases. However, improved coordination between irrigation authorities, farmers, and laborers will be necessary to obtain these benefits and to avoid the negative consequences of excessive synchronization.

An integrated program of pest control might include as well the planting of short- to medium-duration varieties, which would, as pointed out earlier, contribute to pest suppression. Durable varietal resistance would augment the effects of reduced availability of rice to its pests. There is a lack of consensus, however, regarding the types of resistance and the means of deploying them that would favor durability. It is apparent nonetheless that monogenic resistance, deployed in uniform varieties over wide areas, is not a viable long-term strategy at least for certain pests. Polygenic or multigenic inheritance may prove more difficult for pests to overcome and there is evidence as well that moderate resistance may help to conserve natural enemies by maintaining low pest numbers.

Although the ultimate limitation to the densities of rice pests appear to be set by host plant availability, natural enemies play a major role in their dynamics. This is suggested both by the evidence presented above which shows that the seasonal growth rate of pest populations is density dependent and by detailed investigations in the field of the response of parasites and predators to the numbers of specific pests. There are indications that pesticide-induced resurgence of brown planthopper, green leafhopper, and leaffolder was involved in the outbreaks of these pests in the early and mid-1970's, but we suggest that devastating populations could only build up in agroecosystems in which the duration of the rice host had been extended by double cropping and asynchrony.

The limitation of host plant availability that we are proposing will, inevitably, constrain natural enemy populations as it does those of their prey. We suggest that a decision is required, whether to favor modification of rice agroecosystems that will hinder pests or encourage their predators and parasites. The evidence available indicates that in rice crops of 3-4 months' duration, the control exerted by natural enemies is insufficient to prevent greater initial numbers at the beginning of a season translating into larger peak densities. We believe that efforts should be directed as far as is practicable to recovering the strict fallow, the "tropical winter" characteristic of traditional rice farming, a fallow that, for maximum effect, should be coordinated across neighboring fields.

References Cited

Anonymous. 1977. Multilines for brown planthopper control. pp. 60-61. In Annual Report for 1976. IRRI, Los Banos, Philippines. 418 p.

Anonymous. 1979. Brown planthopper: threat to rice production in Asia. IRRI, Los Banos, Philippines. 369 p.

Anonymous. 1982. Evolution of the gene rotation concept for rice blast control. A compilation of 10 research papers. IRRI, Los Banos, Philippines. 130 p.

Barker, R. and R. W. Herdt. 1985. The rice economy of Asia. Resources for the Future. Washington D.C., USA.

Chandler, R. F. Jr. 1982. An adventure in applied science: a history of the International Rice Research Institute. IRRI, Los Banos, Philippines. 233 p.

Chang, T. T. 1976. The origin, evaluation, cultivation, and diversification of Asian and African rices. Euphytica 25:425-441.

Chin, K. M. and A. N. Husin. 1982. Rice variety mixtures in disease control. International Conference on Plant Protection in the Tropics. March 1-4, 1982, Kuala Lumpur, Malaysia. 6 p.

De Datta, S. K. 1981. Principles and practices of rice production. John Wiley and Sons, New York. 618 p.

Dyck, V. A. and B. Thomas. 1979. The brown planthopper problem. pp. 3-17. In Brown planthopper: threat to rice production in Asia. IRRI, Los Banos, Philippines. 369 p.

Geertz, C. 1963. Agricultural involution. University of California Press. Berkeley, USA.

Heinrichs, E. A., F. G. Medrano, H. R. Rapusas, C. Vega, E. Medina, A. Romena, V. Viajante, L. Sunio, I. Domingo, and E. Camanag. 1985. Insect pest resistance of IR5-IR62. Intl. Rice Res. Newsl. 10(6):12-13.

Kenmore, P. E., F. O. Carino, C. A. Perez, V. A. Dyck, and A. P. Gutierrez. 1984. Population regulation of the rice brown planthopper (*Nilaparvata lugens* Stål) within rice fields in the Philippines. J. Pl. Prot. Trop. 1:19-37.

Kenmore, P. E., J. A. Litsinger, J. P. Bandong, A. C. Santiago, and M. M. Salac. 1987. Philippine rice farmers' insect control: thirty years of growing dependency and new options for change through training in integrated pest management. *In* Tait, J., and B. Namopeth (eds.) Management of Pests and Pesticides: Farmers' Perceptions and Practices, Westview Press, Boulder, CO.

Kiritani, K. 1979. Pest management in rice. Annu. Rev. Entomol. 24:279-312.

Litsinger, J. A., B. L. Canapi, J. P. Bandong, C. G. dela Cruz, R. F. Apostol, P. C. Pantua, M. D. Lumaban, A. L. Alviola III, R. Raymundo, E. M. Libetario, and R. C. Joshi. 1987. Rice crop loss from insect pests in wetland and dryland environments of Asia with emphasis on the Philippines. Insect Sci. Appl. (in press).

Litsinger, J. A., Price, E. C., Herrera, R. T. 1980. Small farmer pest control practices for rainfed rice, corn and grain legumes in three Philippine provinces. Philipp. Entomol. 4:65-86.

Loevinshohn, M. E. 1984. The ecology and control of rice pests in relation to the intensity and synchrony of cultivation. Ph.D. thesis, Centre for Environmental Technology, University of London. 358 p.

Manwan, I., S. Sama, and S. A. Rizvi. 1985. Use of varietal rotation in the management of tungro disease in Indonesia. International Rice Research Conference, 1-5 June 1985, IRRI, Los Banos, Philippines. 19 p.

Painter, R. H. 1951. Insect resistance in crop plants. MacMillan and Co., New York.

Reissig, W. H., E. A. Heinrichs, and S. L. Valencia. 1982. Effects of insecticides on *Nilaparvata lugens* and its predators: spiders, *Microvelia atrolineata* and *Cyrtorhinus lividipennis*. Environ. Entomol. 11:193-199.

Sogawa, K. 1976. Rice tungro virus and its vectors in tropical Asia. Rev. Pl. Prot. Res. 9:21-46.

Vergara, B. S. and T. T. Chang. 1976. The flowering response of the rice plant to photoperiod. A review of the literature. IRRI, Los Banos, Philippines. 75 p.

Yoshida, S. 1981. Fundamentals of rice crop science. IRRI, Los Banos, Philippines. 269 p.

Zandstra, H. G., E. C. Price, J. A. Litsinger, and R. A. Morris. 1981. A methodology for on-farm cropping systems research. IRRI, Los Banos, Philippines. 149 p.

9

Entomology of Wheat

M. J. Way

INTRODUCTION

The world-wide distribution and the area devoted to wheat exceed that of any other arable crop (Fig. 1: Table 1). Wheat is more nutritious than other major cereals, it provides more nourishment than any other food source and enters more into international trade (Briggle 1980, Harlan 1980).

Primitive wheats and barleys supported the earliest known agriculturalists about 11 thousand years ago in Syria. Wheat was grown in the Nile valley before about 5000 B.C. and reached India, China and Western Europe around 3000 B.C. It was first grown in Mexico in A.D. 1529, in the U.S.A. in 1602 (an island off the coast of Massachusetts), and in Australia in 1788. African 'tropical' wheats were introduced to the Highlands of Kenya in the late 1800's. The most important recent developments have been the application of high yield technology, especially in N.W Europe since the mid 1950's, and Green Revolution wheats in developing countries such as Mexico and the Indian sub-continent.

In this paper, the range of insect pests attacking wheat in different regions is surveyed and the phenology of chosen species discussed in relation to that of the crop. In particular, the methods that can decrease pest damage to wheat are considered and particular case histories are used to show how they can be manipulated in different pest, climatic, agricultural, social and economic conditions.

THE ORIGIN OF MODERN WHEATS

Much is known about the evolution of wheat in the Near/Middle East from diploid species such as the wild and cultivated Einkorns *Triticum monococcum* through tetraploids such as wild and cultivated

Table 1. - Wheat production in the year 1983 in the main cropping regions. (FAO Production Yearbook)

	Area harvested 000,000 h.	Production 000,000 t.	Mean yield t./h.	Mean yield range t./h.
WORLD	230	498	2.16	0.36, Somalia - 7.04, Netherlands
N. America	39	76	2.41	1.97, Canada - 2.65, U.S.A.
S. America	10	16	1.58	0.57, Bolivia - 1.71, Argentina
Asia (exc. U.S.R.R.)	82	171	2.08	0.83, Iraq - 4.22, S. Korea
Western Europe	19	74	3.93	0.72, Portugal - 7.04, Netherlands
U.S.S.R.	51	82	1.61	
Australasia	13	22	1.73	1.72, Australia - 3.75, New Zealand

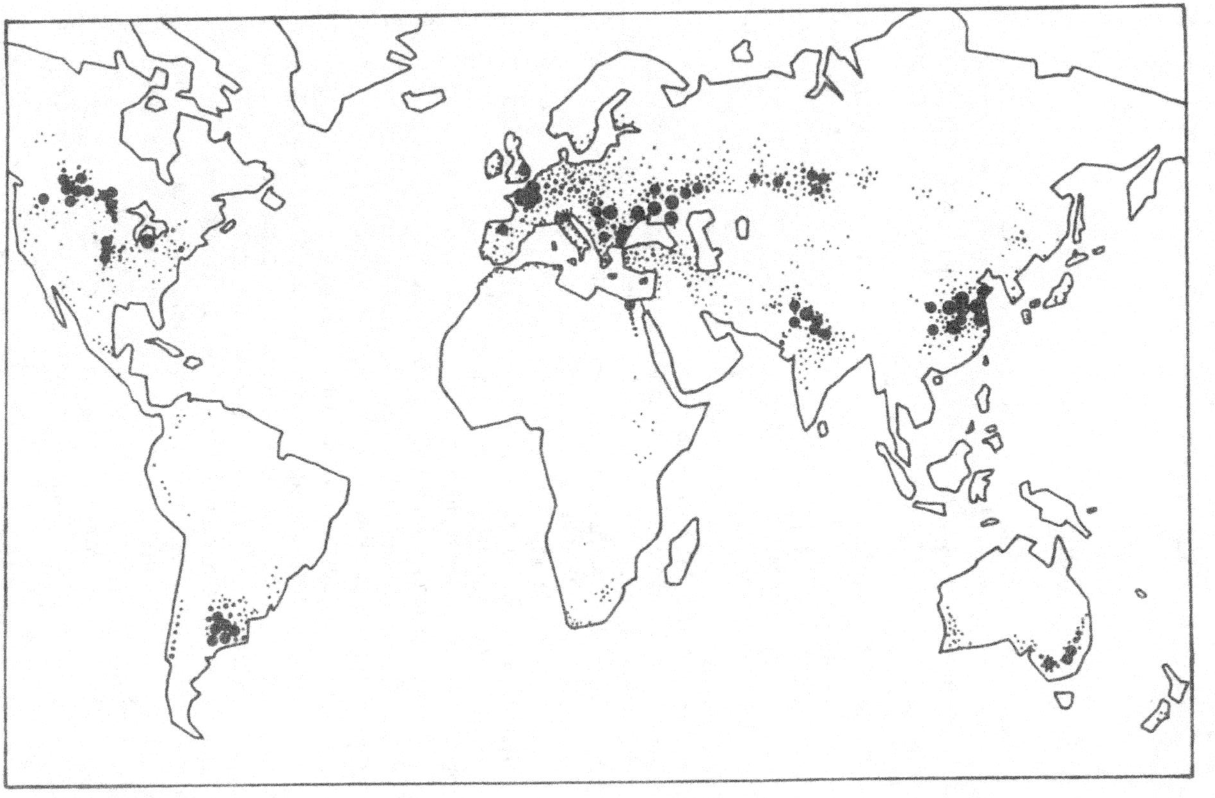

Fig. 1. World wheat production areas (Briggle 1980).

Emmer and Durum (macaroni wheat) *T. dicoccum* and *T. durum*, to the domesticated hexaploids, notably bread wheat *T. aestivum* (Briggle 1980, Harlan and Zohary 1966, Zohary 1969). Wild wheats still occur commonly in parts of the Near/Middle East, sometimes as virtual monocultures.

CLIMATIC LIMITS FOR WHEAT PRODUCTION

The area of origin of wheat in the Near/Middle East is predominantly semi-arid with mild moist winters and hot dry summers, the latter preventing the establishment of natural plant succession beyond grassland or savannah. Grasses, including wild wheats, are therefore a major component of the natural climax vegetation.

Most of the world's wheat is now produced in semi-arid and sub-humid conditions (25-75 cm rain per annum) though some is also grown in irrigated arid regions (Peterson 1965). The places where wheat is now mainly cultivated (Fig. 1, Table 1) have climates as follows: (a) similar to the area of origin - e.g. dry Mediterranean conditions of Europe, California, South America and Australia; (b) the Great Plains of North America where hot, relatively very dry summers and, also very cold winters in the north maintain prairie or savannah vegetation as a dominant part of the climax vegetation; (c) much of the U.S.S.R., with a climatic range and natural vegetation similar to that of the Great Plains but with less stable rainfall; (d) mostly arid regions with mild/hot dry 'winters' and very hot dry summers where wheat is grown under irrigation in winter e.g. the west coastal plain of Mexico, most wheat areas in the Indian sub-continent, and the Sudan; (e) maritime regions with a cool moist winter and a warm/hot, usually moist summer - e.g. north-west Europe and coastal parts of the Pacific north-west of the U.S.A.; (f) high altitude equatorial tropics at about 1500 to 3500 m. with notable diurnal but little seasonal climatic change, and a climax forest vegetation - e.g. parts of Bolivia and Kenya.

Breeding for temperature tolerance has made it possible to grow wheat in climatic extremes varying from the sub-tropics where it is sown in 'autumn' and harvested in 'spring' to within the arctic circle where the crop is spring sown and can mature in the short summer. Climate, however, imposes major constraints on yielding capacity. Spring wheats, with a shorter growing season, are inherently lower yielding, thereby limiting yielding capacity in much of Canada and northern U.S.S.R. where cold winters mostly prevent crops being sown in autumn. Yield potential is limited by high temperatures in the irrigated sub-tropics and also in rain fed systems where the summer is hot and dry as it is in most of the major wheat growing regions. Climate may also limit responsiveness of crops to yield-increasing inputs such as fertilizers. Much the highest yield potential is in temperate climates, notably north-west Europe and also parts of the Pacific north west of the

U.S.A. where the natural climax vegetation is forest in contrast to grassland or savannah in the other main wheat producing regions. As discussed later this has implications for insect host plant relationships, for the damage caused, and for control strategies.

INSECT PESTS

Bonnemaison (1980) and Gallun (1980) give large but incomplete lists of insects that have been recorded as pests of wheat. Many more species feed on wheat but appear to be relatively harmless.

Some wheat damaging species belonging to the following taxonomic groups are polyphagous: - Isoptera, Orthoptera, Elateridae, Scarabaeidae, Noctuidae, Tipulidae. These will be discussed in detail but will be considered under the general heading of pest management. Emphasis will be given to species that are oligophagous for Graminae, notably from the selected list of important pests of wheat in Table 2. All are associated with native wild grasses, including wild wheats in the area of origin. The guild of pests associated with wheat in its origin will be discussed and contrasted with the range of species that are important in some of the regions where wheat is now grown. In particular, relevant characteristics of the ecology of selected pests are outlined as a basis for their management. Aphids and the Diptera are highlighted as much the most important insect pest groups in the regions where most wheat is now grown.

Near/Middle East & associated regions of Southern Europe, Central Asia and North Africa

A unique guild of important pests occurs in this area which includes the area of origin of wheat. Some species appear to be localized or are common only in parts of the Near/Middle East.

Important pests include the following:-

Margarodes tritici (Homoptera, Margarodidae). This scale insect attacks grass weeds as well as being a locally serious pest of wheat in Turkey (Duran 1976). The adult feeds on below-ground stems and roots between April and July. The eggs "incubate" (? diapause) for about four months in the adult "cocoon", followed by newly hatched larvae which also diapause in the cocoon from September until about the following May. Larvae actively feed between January and June. The double diapause is an adaptation to the long hot dry off-season; also non-dispersiveness is associated with the 'permanence' of annually recurring wild grasses and wheats; pest status is linked with continuous cereal production.

Eurygaster spp. notably *integriceps*, and *Aelia* spp. (Heteroptera, Pentatomidae). Pentatomidae are strikingly associated with cereals and wild grasses in this region. There is remarkable diversity, with some 37

TABLE 2. Some important wheat pests that are oligophagous for graminae, and their approximate world distribution on wheat.

HOMOPTERA
  Aphididae
      Rhopalosiphum padi L., Bird Cherry Aphid. Virtually world-wide.
      R. maidis Fitch, Cereal Leaf Aphid. Virtually world-wide except where winters are severe.
      Schizaphis graminum Rond., Greenbug. Southern Europe, Near/Middle East, Southern & Central Asia, S. Africa, N., Central & S. America (not in colder latitudes).
      Anuraphis maidiradicis Forb., Corn Root aphid.
      Diuraphis tritici Gillette, Western Wheat Aphid. N. America (Colorado, Montana, Illinois).
      D. noxius Mordv. S. Europe, Near/Middle East, Central Asia, N. & S. Africa, Argentina.
      Metapolophium dihrodum Walk., Rose Grain Aphid. Europe, Near/ Middle East, Central Asia, Africa, N. & S. America, Australasia.
      M. festucae Theob., Fescue Aphid. N. Europe, Iceland.
      Sitobion avenae F., Grain Aphid. Europe, Near/Middle East, Central Asia, Indian Region, N., & S. Africa, N., Central & S. America.
      S. fragariae Walk. Europe, Near/Middle East, N. & S. Africa, Australasia, Western N. America.
  Margarodidae
      Margarodes tritici. Near/Middle East.
HETEROPTERA
  Pentatomidae
      Eurygaster integriceps Put., Sunn Pest. Much of Near/Middle East, S. Eastern Europe, Soviet Central Asia.
      E. austriaca Schr., Sunn Pest. Near/Middle East, Eastern Central Asia, N. Africa.
      Aelia rostrata Boh. Near/Middle East.
      E. maura L. Near/Millde East, Eastern Central Asia, N. Africa.
      A. acuminata L., Pointed Wheat Shield bug. Southern Europe, Near/Middle East, Soviet Central Asia, N. Africa.
  Lygaeidae
      Blissus leucopterus Say., Chinch Bug. N. & S. America.

Table 2 (cont.)

Macchiademus (Blissus) diplopterus Dist., Grain Chinch bug. Southern Africa.

COLEOPTERA

Carabidae

Zabrus spp. notably tenebroides, Corn Ground Beetle. S. E. Europe, Near/Middle East, Soviet Central Asia.

Chrysomelidae

Oulema melanopa L., Cereal Leaf Beetle. Europe, parts of Near East, U.S.S.R., N. Africa, N. America.
Oulema erythrodera Lac., Grain Slug. Southern Africa.
Pseudapophylia smaragdipennis, Sandveld Grain Worm. Southern Africa.

Curculionidae

Barytychius (Pachytychius) hordei Brulle. Near East, S. E. Europe, N. Africa.

LEOPIDOPTERA

Scythridae

Syringopais temperatella Led. Near/Middle East.
(see Bonnemaison (1980) for list of many other noteworthy Lepidoptera that attack wheat and other Graminae)

DIPTERA

Cecidomyidae

Contarinia tritici Kirb., Lemon Wheat Blossom Midge.
Sitodiplosis mosellana Geh., Orange Wheat Blossom Midge. Europe, U.S.S.R., U.S.A.
Haplodiplosis marginata Wagn., Saddle Gall Midge. Europe, U.S.S.R.
Mayetiola destructor Say., Hessian Fly. Europe, U.S.S.R., U.S.A.

Opomyzidae

Opomyza florum F. Europe, U.S.S.R.

Chloropidae

Chlorops pumilionis Bjerk, Gout Fly. Europe, U.S.S.R.
Meromyza americana Fitch, Wheat Stem Maggot. U.S.A.
Oscinella frit L., Frit Fly. Europe, U.S.S.R., U.S.A.

Muscidae

Delia coarctata Fall., Wheat Bulb Fly. N. Europe, Western U.S.S.R., (Iraq?)]
Antherigona spp. notably *falcata* and *naquii*

Table 2 (cont.)
HYMENOPTERA
    Cephidae
        Cephus pygmaeus L., Wheat Stem Sawfly. Europe, N. Africa, Near/Middle East, Central Asia.
        C. cinctus Nort., Western Wheat Stem Sawfly. Canada, N. U.S.A.
    Chalcidae
        Harmolita tritici Fitch, Wheat Joint Worm. U.S.S.R., U.S.A. H. grandis, Wheat Straw Worm. U.S.S.R., U.S.A.

---

species identified from Turkey of which 6 belong to the genus *Eurygaster* and 12 to the genus *Aelia* (Lodos 1981 a, b, Paulian and Popov 1980, Talhouk 1969). Only three of the former genus and two of the latter are considered to be economically important, but these are often the most destructive pests in their areas of distribution. *E. integriceps* is most widely damaging and its life cycle is representative, although relatively complex. Egg laying adults occur from about February to May. They and their progeny feed on the shoots while the nymphs, as well as the next generation of adults, feed on the developing grain. They kill the growing point and diminish seed yield and quality. Fully fed adults may migrate in July and August to aestivating/hibernating sites (mostly mountains in the southern part of its range, mostly forests in its northern range). They return to crops in February/March of the following year. The group as a whole exemplifies successful adaptation to the long dry summer as well as individual species diversity e.g. in strategies, some non-migratory, others migratory associated with long-standing co-evolution with wild and cultivated Graminae in particular localities (Brown 1965). Local success as indicated by the severity of attacks by *E. integriceps* may indicate the selective value of migrating about 20 km to 200 km, to and from favorable hibernating sites, though wind direction may influence seasonal pest status (Brown 1965). Nevertheless counts of overwintering adults can sometimes provide a useful forecast of likely damage to crops (Adiguzel 1981).

Barytychius hordei (Coleoptera: Curculionidae). Reproducing adults feed from early April to mid June on leaves, stems and developing grain. The larvae feed on the grain, and pupate in a cocoon in the soil in which the adult diapauses from late July. Some adults emerge in the following year but others remain in diapause for a further year (Koyuncu 1976). The diapause characteristics, together with flightlessness are adaptations to the climate and to a continuous grassland or cereal crop habitat.

Zabrus tenebrioides (Coleoptera: Carabidae). This is the only seriously damaging Carabid species but a many as twenty-five other *Zabrus* species have been identified on wheat in Turkey (Altinayar 1976 b).

Syringopais temperatella (Lepidoptera: Scythridae). Adults lay eggs in soil from April to May. According to Rivnay (1956) the larvae began to hatch in early June, feed for about a week, and then diapause in the soil until rains begin in autumn (Kaya 1976). The larvae mine in the leaves of the young plants mostly pupating in the mines, and the adults begin to emerge in April. This is another species that must depend on continuous 'crops' of Graminae since the eggs are laid before the next season's crop begins to develop. Crop rotation, ploughing and clean fallowing completely control this pest (Harpaz 1963).

The above and other oligophagous species are all recognized as major wheat pests in some localities in the Near/Middle East Region. Although insects are of minor importance in some countries of the Region, e.g. Israel, the overall situation contrasts with other regions where there are usually only two or three species or families that are considered to be severely damaging. Besides the species that are limited to parts of the Near/Middle East, southern Europe, North Africa and central Asia, there are a few Near/Middle East pests that have become much more widely distributed, for example, *Cephus pygmaeus* the wheat stem sawfly, which is the only major sawfly pest among a group of about nine species on wheat in Turkey, the other being much more localized in distribution (Altinayer 1976 a). Other species which occur elsewhere include *Mayetiola destructor*, the Hessian Fly, *Oscinella frit* the Frit Fly and *Lema melanopa* the Cereal Leaf Beetle which appear to do little harm in the Near/Middle East. Lodos (1973) gives a useful list of the major wheat pests in Turkey, which is probably relevant to some other Near/Middle East countries. It is significant that this list does not include aphids or Diptera.

Pests in Western Europe

The major pest species (Table 3) contrast strikingly with those in the Middle East. Unlike most of the Middle East and the prairie regions in North America and Australia, the natural climax vegetation is forest. Yet, the pests are seemingly indigenous species which must have survived before-agriculture on wild grasses, sparsely distributed in places where they could survive, notably after landslides and floods bared the ground. Here, there would be colonization by ephemeral grasses and other herbaceous plants with which appropriately adapted insects and other organisms are associated. For example, some aphids species are supremely well adapted to seeing green plants silhouetted against a bare soil background as occurs in nature with early plant colonists dispersed on recently exposed bare soil. Crop plants are similarly highly attractive, at any rate during early stages of growth.
The wheat bulb fly *Delia coarctata* is a striking example of adaptation to the earlier stages of plant succession since the eggs are laid in August on bare soil at a time when there is no host plant attractant, the survival of the species on a particular piece of land being dependent upon subsequent germination of wild host grasses or wheat which is normally

Table 3. Important pests of wheat in western europe.

HOMOPTERA (Aphidae)
     Rhopalosiphum padi - bird-cherry aphid
     Schizaphis graminum - greenbug
     Metopolophium dirhodum - rose grain aphid
     Sitobion avenae - English grain aphid
DIPTERA
     Contarinia tritici - wheat blossom midge
     Sitodiplosis mosellana - wheat blossom midge
     Chlorops pumilionis - gout fly
     Delia coarctata - wheat bulb fly

sown after mid September. It is remarkable that this species lays no eggs if the soil is well covered by vegetation in August - so it is unable to colonize solid grassland or wheat monocultures where the wheat remains unharvested and the land unploughed until September. The implications for control of such insects will be discussed later in this chapter.

<u>Pests</u> <u>of</u> <u>North</u> <u>America</u>

North American wheat is grown in ecological conditions ranging from those like the Middle East to those in Western Europe. The limited range of major pests (Table 4) therefore includes species such as the wheat stem sawfly, *Cephus cinctus* with a life cycle and ecological requirements essentially similar to that of *C. pygmaeus* in the Middle *East*. The cereal leaf beetle, *Oulema melanopa* and Hessian Fly, *Mayetiola destructor* were both introduced into North America from Europe or Eurasia. The aphids may also have originated in Europe. So, there is a mostly introduced population of wheat pests with the potential for further introductions and further spread. For example, the Hessian Fly, first recorded in 1778 in the U.S.A. did not reach Washington State until 1960, and Texas in 1978 and is still spreading. The situation is similar in South Africa where, for example, the Russian wheat aphid *D. noxia* has become an important pest since its appearance in 1978. The first occurrence of *Metopolophium dirhodum* in Australia in 1983 indicates dangers for that country, which, being ecologically isolated, has so far been comparatively free from wheat insect pest problems.

CASE HISTORIES

<u>Aphidae</u>

About ten species of aphids are recognized as direct pests of wheat and sometimes also as vectors of virus diseases (Table 1). *Rhopalosiphum*

Table 4. Important pests of wheat in North America.

HOMOPTERA
Schizaphis graminum - greenbug
Sitobion avenae - English grain aphid
COLEOPTERA
Oulema melanopa - cereal leaf beetle
DIPTERA
Mayetiola destructor - hessian fly
HYMENOPTERA
Cephus cinctus - wheat stem sawfly

spp are virtually worldwide in distribution, some such as *Shizaphis graminum* and *Diuraphis noxius* are rare or absent in cooler latitudes. Some have not yet reached their potential distribution, and represent a serious quarantine problem for Australia in particular (Table 2).

Cereal aphid life cycles are typical of Aphidae in general. Most species have races which are holocyclic in colder conditions but solely anholocyclic where winters are warm. In intermediate regions such as north-west Europe both races coincide. Where holocycly occurs, some reproduce both asexually and sexually on Graminaceous hosts and others, although limited to Graminae in the parthenogenetic phase reproduce sexually and overwinter on dicotyledonous hosts (Blackman and Eastop 1984).

The pest status of aphids in general varies strikingly in different regions. As already mentioned, they are not recognized as pests in the area of origin of wheat and seemingly cause little or no harm even in conditions where some wheat is grown fairly intensely, as in Israel. Yet, in similar climates in southern U.S.A. and Argentina, the green bug *S. graminum*, for example, can be a serious pest. Most aphids cannot survive high temperature, and this combined with dry summer conditions presents a serious challenge to continuous survival in hotter regions even when winter and spring conditions favour aphid multiplication. Long distance migration, as in U.S.A., and powers of multiplication are the basis for aphid success, particularly where hot dry conditions may seasonally eliminate a species. Perhaps differences in natural enemy action explain some of the anomalies e.g. the severity of *S. graminum* in parts of southern U.S.A. whereas in South America, parasite introductions appear to have alleviated some aphid problems.

There is no doubt that a crucially important factor influencing the success of aphids and their importance as pests of wheat and other crops is the crop production technology that is adopted. This is well exemplified by developments in northwest Europe during the last 20 years. In the U.K., for example, there were occasional aphid upsurges on wheat before the 1960's but seemingly not sufficient to justify control. This contrasts strikingly with the present status of aphids as the most serious pests of wheat. Their development as wheat pests is primarily associated with breeding for yield at the expense of various forms of resistance (Way and Cammell 1979) and with large increases in application of nitrogenous fertilizers, together with changes in crop practices, for example, early sowing of winter wheat which becomes colonized by viruliferous aphids soon after sowing in September or early October, contrasting with the earlier practice of sowing the crop after about mid October when aphid migration had ceased. Fig. 2a indicates that the economic injury threshold is lowered as the yield expectation is increased. Fig. 2b shows that nitrogen fertilizer, which not only makes plants more attractive to colonizing aphids, also increases its multiplication rate once a population is established. Other yield-increasing changes have affected natural enemy action, first by making natural enemies less able to combat the quicker multiplying aphids and secondly, by harming natural enemies directly by pesticides and indirectly through changes towards pure monocultures (Sunderland et al. 1985). At least 390 species of non-specific predators are recorded from U.K. cereals. Many prey on aphids but much research suggests that their action is unpredictable though undoubtedly in some cereal fields at some times important biological control occurs (Sunderland et al. 1985).

Much emphasis has been placed on host plant resistance against wheat aphids, in particular the green bug *Schizaphis graminum* (Gardenhire 1980), but seemingly, good resistant cultivars are not yet available as commercial varieties. The three components of inherent resistance, namely non-preference (antixenosis), antibiosis and tolerance have been identified in the laboratory from different sources though resistance may vary according to aphid biotype.

Lowe (1982) has suggested that serious outbreaks of aphids in the U.K. in the 1970's were associated with widespread use of two cultivars which were particularly susceptible compared with varieties used before and afterwards. Nevertheless, the resistance in all modern varieties in Europe would seem to be relatively small. Resistance in wheat to aphids is highly complex with different forms of resistance and with variation according to plant age. Resistance may also differ against different aphids, and depend on the previous host of the aphid as well as the aphid biotype. It is also possible that 'resistance' may be influenced by the attractiveness of a particular cultivar to natural enemies particularly some non-specific ground predators (Sunderland et al. 1988).

The Future of Aphid Control on Wheat and Other Small Grain Cereals

High yield technology as practised in northwest Europe is seriously jeopardizing controls based on methods other than insecticides. Furthermore, low economic thresholds make it impossible to apply aphicides other than as a routine. In these circumstances the key

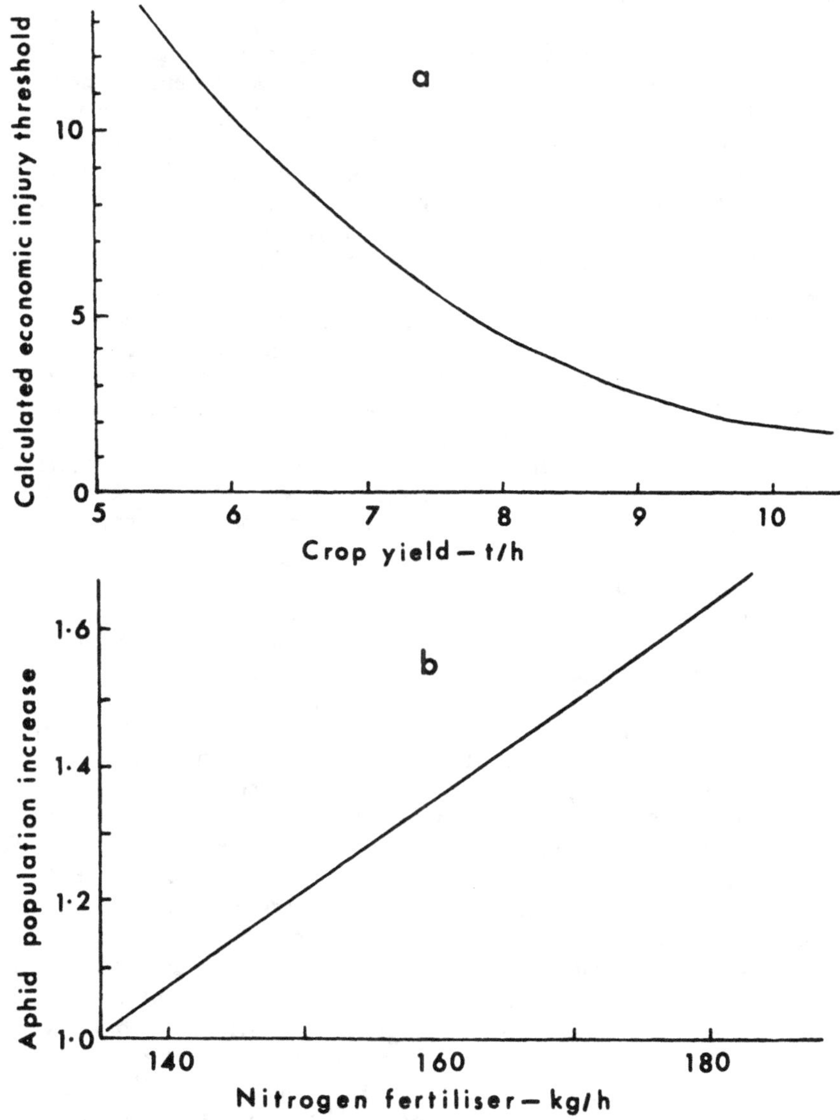

Fig. 2. Change in economic injury threshold in relation to potential crop yield (a) and aphid population increase in relation to nitrogen fertiliser (b). Rabbinge et al., 1983; Vereijken, 1979.

alternative to insecticides is host plant resistance, although premiums on quality and yield have so far made it impossible to produce high yielding cultivars without trading yield for required resistance.

Where the yield objective is lower, a larger and measurable aphid population can be tolerated before imposed controls such as insecticides are needed. Moreover natural enemies can perhaps be encouraged by cultivation practices such as undersowing as well as by use of relatively selective insecticides such as pirimicarb; also, in these circumstances host plant resistance also becomes more feasible even at the expense of some yield potential. In particular, small amounts of resistance can complement natural enemy action and also make aphids more susceptible to insecticides, thereby permitting fewer and weaker applications (Starks et al. 1970, van Emden 1983). Less nitrogen fertilizer also increases host plant resistance and it becomes more feasible to manipulate the crop to escape aphids in time and space. Thus, in northwest Europe escape in time by delayed sowing, as already indicated, can greatly decrease damage by aphid transmitted virus diseases. Escape in space is less easy against species that are so highly adapted to find and colonize new hosts, but undersowing, for example can reduce the ability of some important aphids to find their host plants - a practice which camouflages the otherwise attractive clear pattern of green plants silhouetted against bare soil.

<u>Wheat Stem Sawfly - *Cephus cinctus*</u>

The wheat stem sawfly in the North American prairie's and *C. pygmaeus* in Eurasia (also in parts of eastern U.S.A.) are striking examples of species adapted to climax vegetation grassland. Their limited powers of dispersal and habit of colonizing field edges means that they do little harm where wheat is part of a rotational system, and conversely can be serious pests of monoculture wheat as in North American prairie's and parts of the Near/Middle East. The pest lays eggs within the stems of young wheat plants, the larva develops in the stem and just before completing development 'cuts' the inside of the base of the stem and forms a cocoon in the 'stub' of the wheat plant. The stem soon breaks off at the point of the 'cut'. The insect dispauses as a larva in the 'stub' and the adult emerges the following spring.

Table 5 indicates the changing status of American wheat stem sawfly in the Canadian prairie's (Holmes 1982, Morrison unpublished, Turnock 1977). In the wild pre-agricultural situation there was presumably a balanced population undistributed by the earliest scattered wheat crops in which ploughing in of stubble also killed many diapausing larvae. However, after about 1910, changes mostly favoured the sawfly which became a serious pest until after 1955 when resistant cultivars and other factors seemingly caused decline in severity. The sawfly example shows how ploughing, crop rotation, and resistant cultivars (even though at least 30% of sawfly survive in such cultivars (Holmes 1982), and opportunities for increased parasitism prevent wheat stem sawfly becoming a pest. This has also been demonstrated by wheat

Table 5. Possible causes of changing status of wheat stem sawfly in S. Canadian Prairies.
(R. J. MORRISON, unpub. and Turnock, 1977)

(1) <u>Virgin prairie</u> (population in 'balance')
 (a) Well distributed and abundant western wheat grass wild host.
 (b) Always almost half of stems suitable for oviposition.
 (c) Soil undisturbed so diapausing larvae unharmed.
 (d) Stable situation for parasite/host equilibrium except prairie fires killed parasites.
(2) <u>Cropping before c. 1910</u> (sawfly unimportant)
 (a) Large dispersed wheat field - only edges attacked.
 (b) Late maturing cultivars allowed parasite build up.
 (c) Stem rust epidemics obliterated sawflies.
 (d) Rotations with non-host oats hampered pest build up.
 (e) Ploughing buried diapausing larvae.
 (f) No summer fallow before 1885 so sawfly limited by moisture.
(3) <u>Later wheat farming</u> (serious pest)
 (a) Change in preference of sawfly from native grasses to wheat.
 (b) Extensive monocropped land allowed pest build-up.
 (c) Strip cropping increased 'edge' attack.
 (d) Summer fallowing produced excellent wheat stems for oviposition and larval development.
 (e) Early maturing cultivars decreased parasitism.
 (f) Stem rust resistance benefited sawfly larvae.
 (g) Stubble burning killed parasites.
 (h) Mechanisation reduced rotations with oat crops previously needed for horses.
 (i) Change from ploughing to surface cultivations did much less harm to diapausing larvae.
(4) <u>Post 1955</u> (sawfly decline)
 (a) Resistant cultivars.
 (b) Certain host grasses sown by roads and ditches are highly suitable for parasites.
 (c) Sawfly source in western wheat grass prairie reduced.
 (d) Increased field width decreased 'edge' attack.
 (e) General climatic changes - less rain and higher temperatures in summer unfavorable for sawflies.
(5) <u>The future</u> (some of the likely changes potentially favorable to sawfly)
 (a) Zero tillage ideal for maximum sawfly survival.
 (b) Summer fallow replaced by alternatives (grass strip borders, zero tillage) highly favorable to sawfly.
 (c) New later maturing cultivars will favor sawfly parasites.

production in much of Israel where *C. pygmaeus* creates no problems compared with its status as a serious pest in parts of the Near/Middle East where wheat is grown traditionally as a monoculture using simple, non ploughing, cultivation techniques. Other pests Near/Middle East species with a comparable life cycle are similarly controlled by such practices in Israel (Harpaz 1963).

<u>Diptera - Wheat Bulb Fly and Hessian Fly</u>

The life history of wheat bulb fly is strikingly adapted to colonization of early stages in succession of certain grasses. Wheat grown to coincide with the usual germination time of wild host grasses can be catastrophically attacked if the soil is bare and has therefore attracted adults during the oviposition period in August. Fig. 3 indicates strategies for combating the pest based primarily on knowledge of pest biology (Anon 1985, Long 1960). Until recently the favoured practise was to sow 'winter' wheat in October/November, so coinciding with the susceptible stage of the crop because by February or March it may not have grown sufficiently to compensate for destruction of tillers by the larvae. Now, many winter wheat cultivars are being sown earlier in September or even late August. In these conditions the strong plants can tolerate damage in February-March without loss of crop, but as indicated by Fig. 4 the crop is then subject to attack by several others pests that are active in August-September but not October, in particular viruliferous aphids. Fig. 4 shows that a sequence of methods can be used to control wheat bulb fly, so it would seem better to risk wheat bulb fly attack than the more difficult-to-control virus diseases.

The Hessian fly, a serious pest in parts of the U.S.A. and sometimes in central Europe has a much more complex life cycle than that of the wheat bulb fly, with two or more generations a year attacking both young and old plants by feeding on the stems inside the leaf sheaths (Fig. 4). Like wheat bulb fly, sowing date maybe of crucial importance in northern parts of the U.S.A. For example, in north central states, records of adult activity have provided long standing advice on safe sowing dates - after about mid-September in northern Illinois and after about mid-October in southern Illinois (Metcalf and Flint 1962). However, during the last 10-20 years the fly has spread into the west and south and the U.S.A. where the winter is not sufficiently cold to create a significant fly free period during which wheat can be safely sown. There has been considerable success in some parts of the U.S.A. in developing resistant cultivars (Gallun 1980). Unfortunately such resistance is not yet compatible with other crop requirements for much of the U.S.A., including regions such as Texas where there is a very limited fly free period (Hoelscher and Turney 1985). Control tactics for north central states of the U.S.A. are indicated in Fig. 5. In the southern states, control may depend on ploughing in of stubble immediately after harvest, which destroys many pupated insects, and on destruction of alternative hosts such as volunteer wheat. The occurrence of at least nine biotypes of Hessian fly (Gallun 1980) each requiring different resistance

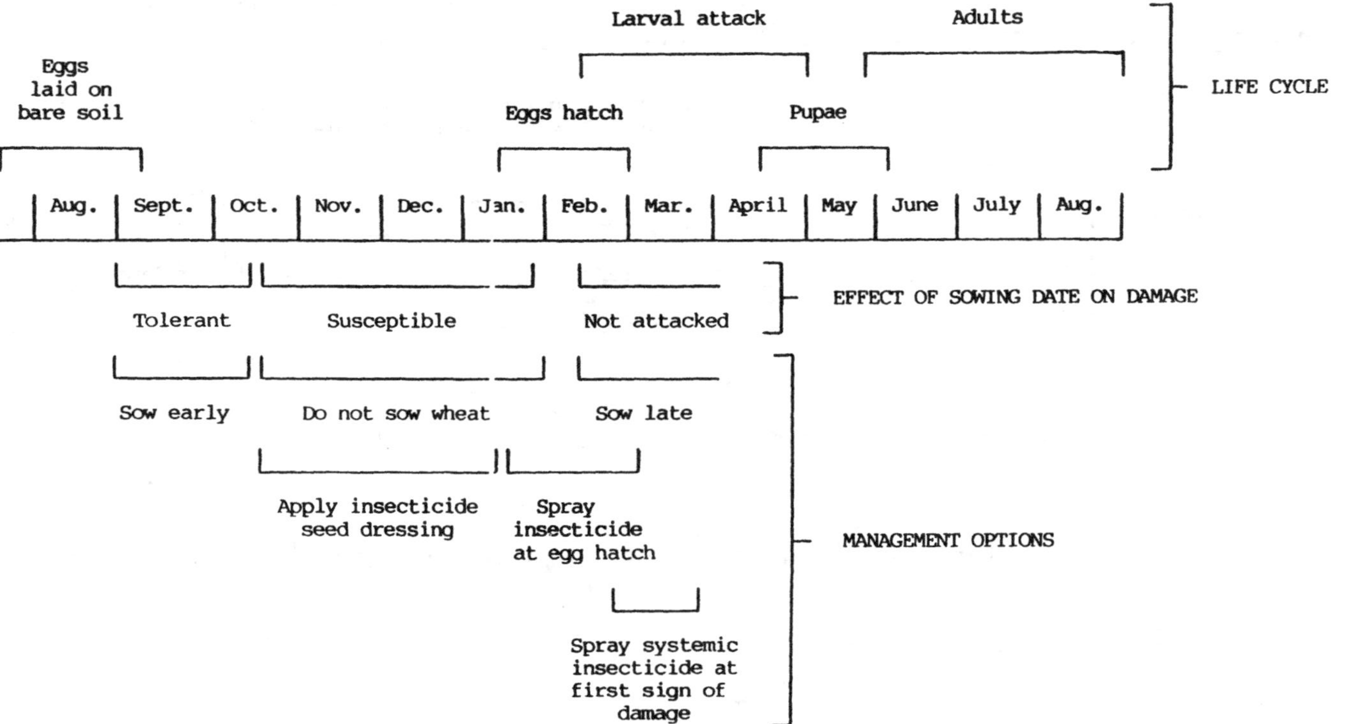

Fig. 3. Management of Wheat Bulb Fly, _Delia coarctata_ in relation to life cycle and sowing date of wheat in the UK.

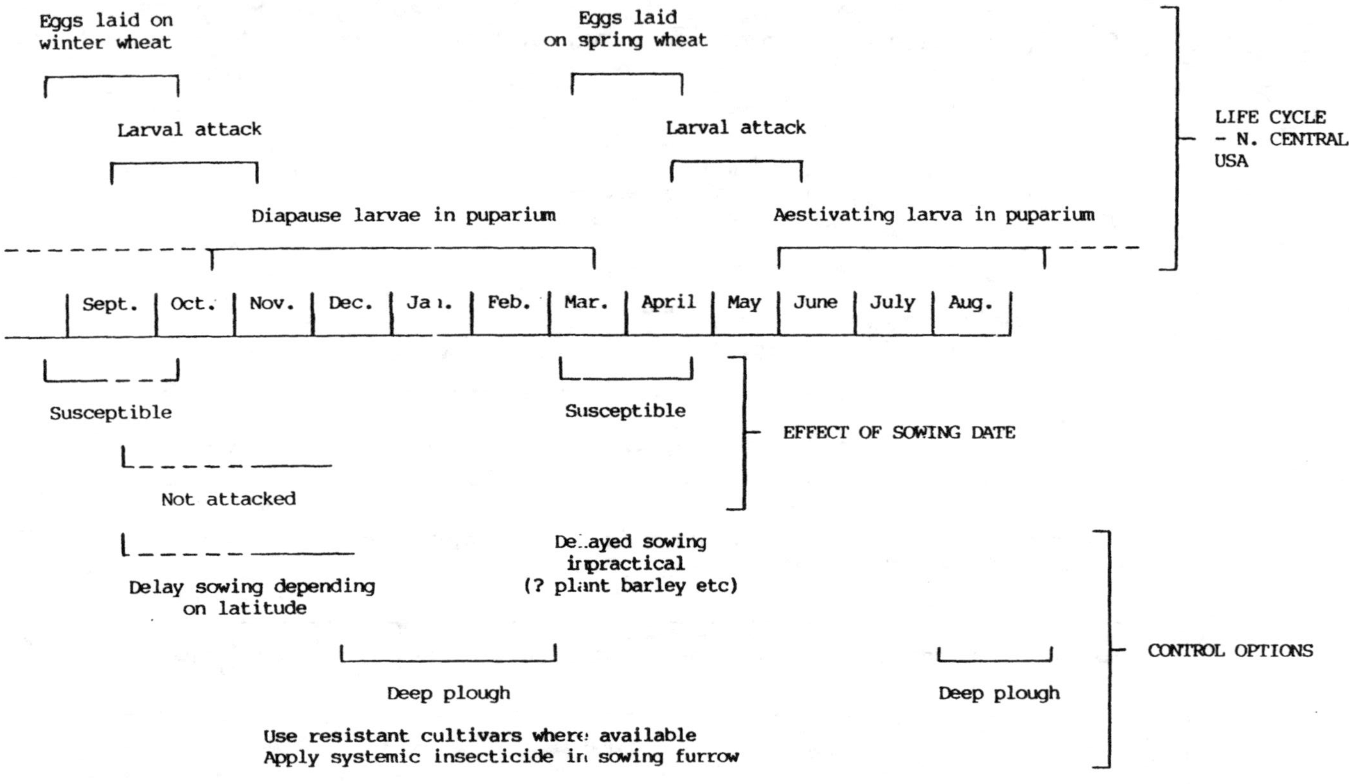

Fig. 4. Management of the Hessian Fly *Mayetiola* *destructor* in relation to life cycle on wheat in North Central USA.

characteristics in wheat adds to the difficulty of controlling this pest which is perhaps the most serious insect pest of wheat in the U.S.A.

Hessian fly, like other comparable pests found in the Middle East, was no doubt originally a denizen of climax grassland but, unlike the other species, its adults are relatively mobile and can be windborne for several miles. Whilst ploughing can be locally effective and crop rotation is helpful, individual fields or groups of fields are not isolated from damaging adult immigrations. Chemical control is also difficult and probably relatively uneconomic for crops grown in much of the U.S.A. where yields are low compared with high input high yield cropping in northern Europe.

DISCUSSION

As already discussed, wheat is grown in a very wide range of conditions from primitive, often subsistence-type agriculture in parts of the Near/Middle East and elsewhere, to high technology 'extensive' production as in much of North America and Australia, or high technology 'intensive' production, notably in northern Europe. A key question is how can insect pest management be improved in the different conditions in the context of different pest ecologies and relevant methods of natural and artificial control?

Natural methods are usually classed as cultural practices, host plant resistance and biological controls. Artificial methods comprise physical controls and chemical pesticides. The flexibility of cropping of wheat and its ease of manipulation in breeding programmes makes cultural controls and host plant resistance potentially very powerful weapons in insect pest control, and the variety of opportunities provided by these two methods is challengingly discussed by Harris (1980). How do the various methods apply economically as well as ecologically to the different crop conditions? As already indicated, pest problems in the area of origin of wheat (Near/Middle East) are largely associated with relatively primitive agriculture with crops grown in conditions ecologically similar to the natural wild wheat grassland but exaggerating features that make the plants more susceptible. Yet, in Israel, for example, most wheat suffers virtually no problems from endogenous pests that can be so serious elsewhere in the Near/Middle East because of the adoption of cultural and physical controls such as manipulation of planting date, crop rotation and ploughing that can severely dislocate the pests' life cycle.

Some or most of the endogenous Near/Middle East pests can be controlled chemically but this is uneconomic in low technology systems. However, the migratory pentatomids are strikingly different from the other pest species because they are not confined to the area of the crop. They can cause catastrophe over fairly large areas but their limitations as exploiters of a crop such as wheat are demonstrated by evidence that area-wide chemical control on crops of these single generation pests can decrease numbers such that next years generation is too small to cause significant damage. Chemical control would therefore seem to be a highly effective and justifiable weapon against insect pests with such characteristics, as it is for locusts, for example.

The opposite extreme is represented by conditions in much of northern Europe where wheat is an introduced plant and the emphasis is on maximizing yield in the face of two important groups of indigenous pests, dipterous stem borers and aphids. Although, in theory, there are outstanding opportunities for pest control by diversification in time and space, the demand for high yields constrains such opportunities as well as necessitating fertilization and high yielding cultivars that also make most insect pest problems worse (Way and Cammell 1979). So, high yield technology has created absolute dependence on virtually routine pesticide treatments and has made other controls uneconomic (Vereijken et al. 1985). This represents a dangerous situation though future constraints, notably decreased European Community support, may lead to much more rational control practices (Way 1978, Way and Cammell 1979). Research developments on experimental farms in the Netherlands are demonstrating relevant integrated pest management practices (Vereijken 1979, Vereijken 1986).

It is in North America, particularly in many parts of the U.S.A., where there seems to be the most rational balance between yield objectives and those of pest management. Although this is partly because climatic conditions usually do not permit intensive wheat production, it is also the result of special emphasis, for example on host plant resistance to some key pests. Table 6 shows that much progress has been made in selecting resistant cultivars although resistance may sometimes be insufficient as a sole means of control or incompatible with other cultivar requirements. Resistance mechanisms are being studied but not used significantly in practice in northern Europe. In both conditions there is great potential for partial resistance which is much more readily obtainable without loss of other good characteristics and should be seen as part of an IPM approach in combination with chemicals and biological controls. Cultural controls, notably escape in time (e.g. against Hessian fly in parts of the U.S.A.) and escape in space (e.g. against wheat stem sawfly) remain as a crucial and fundamental component of control practices in North America, as do physical controls based on ploughing in.

It is significant that the great majority of the world's wheat crop is grown in conditions where combinations of integrated controls are economically feasible, namely in the Americas, Russia and China. This would also apply to Australia, which so far has remained free from invasion by most of the major wheat pests.

Table 6. Current status of intrinsic resistance in wheat cultivars to major insect pests in N. America.

STATUS OF RESISTANT CULTIVARS	
APHIDAE	
Grain aphid, *S. graminum* and others	Generally inadequate as a sole method of control and/or not yet compatible with other required cultivar characteristics. Partial resistance more readily attainable and potentially very valuable in IPM combined with biological control and chemicals.
DIPTERA	
Hessian fly, *M. destructor*	Highly effective in parts of U.S.A. Elsewhere not compatible with other required cultivar characteristics.
HYMENOPTERA	
Wheat stem sawfly, C. cinctus	Relatively effective but crop loss not prevented and yield and some other good qualities traded for resistance.
COLEOPTERA	
Cereal leaf beetle, *O. melanopa*	Seemingly not yet compatible with other required cultivar characteristics.

REFERENCES CITED

Adiguzel, N. 1981. Fluctuations in Sunn-pest populations in Southeastern Anatolia. Bulletin of the European Plant Protection Organization 11: 19-22.

Altinayar, G. 1976a. Cereal stem sawflies in Turkey. In: Anon (ed.) Proceedings of Symposium on Wheat Pests. CENTO Scientific Programme report on. 22, Ankara, pp. 38-47.

Altinayar, G. 1976b. Chemical control of *Zabrus* species in Turkey. In: Anon (ed.) Proceedings of Symposium on Wheat Pests. CENTO Scientific Programme report no. 22, Ankara, pp. 53-56.

Anon 1985. Wheat Bulb Fly. Leaflet no. 177. U.K. Ministry of Agriculture, Fisheries & Food. 7 pp.

Anon 1985. Wheat. A Guide to Varieties From the Plant Breeding Institute. National Seed Development Organization, Cambridge, UK. 80 pp.

Blackman, R. L. and Eastop, V. 1984. Aphids on the World's Crops. Wiley Interscience, Chichester. 466 pp.

Bonnemaison, L. 1980. Principal animal pests. In: Anon (ed.) Wheat. Documenta Ciba-Geigy, Ciba-Geigy, Basle. pp. 59-68.

Briggle, L. W. 1980. Origin and botany of wheat. In: Anon (ed.) Wheat. Documenta Ciba-Geigy, Ciba-Geigy, Basle. pp. 6-13.

Brown, E. S. 1965. Notes on the migration and direction of flight of *Eurygaster* and *Aelia* species (Hemiptera, Pentatomidae) and their possible bearing on invasions of cereal crops. J. Animal Ecology, 34; 93-107.

Duran, M. 1976. The cereal pest, *Margarodes tritici.* In: Anon (ed.) Proceedings of Symposium on Wheat Pests. CENTO Scientific Programme report on. 22, Ankara, pp. 57-64.

Gallun, R. L. 1980. Breeding for resistance to insects in wheat. In: Harris, M. K. (ed.) Biology and Breeding for Resistance to Arthropods and Pathogens in Agricultural Plants. Texas A&M University, College Station, Texas. pp. 245-261.

Gallun, R. L., Starks, K. J. and Guthrie, W. D. 1975. Plant resistance to insects attacking cereals. Ann. Rev. Entomol. 20: 337-57.

Gardenhire, J. H. 1980. Breeding for Greenbug, *Schizaphis graminum* (Rondani) resistance in wheat and other small grains. In: Harris, M. K. (ed.) Biology and Breeding for Resistance to Arthropods and Pathogens in Agricultural Plants. Texas A&M University, College Station, Texas. pp. 237-244.

Harris, M. K. 1980. Arthropod plant interactions related to agriculture, emphasizing host plant resistance. In: Harris, M. K. (ed.) Biology and Breeding for Resistance to Arthropods and Pathogens in Agricultural Plants. Texas A&M University, College Station, Texas. pp. 22-51.

Harlan, J. R. 1980. Origins of agriculture and crop evolution. In: Harris, M.K. (ed.) Biology and Breeding for Resistance to Arthropods and Pathogens in Agricultural Plants. Texas A&M University, College Station, Texas. pp. 1-8.

Harlan, J. R. and Zohary, D. 1966. Distribution of wild wheats and barley. Science 153: 1075-80.

Harpaz, I. 1963. L'emploi de procedes culturaux, plutot que chemiques, dans la lutte contre les parasites animaux et les maladies des cultures dans les regions tropicales pen developpees. J. d'Agriculture Tropicale et de Botanique Applique 10: 603-611.

Hoelscher, C. E. and Turney, H. A. 1985. Hessian Fly in Texas Wheat. Texas Agricultural Extension Service. 6 pp.

Holmes, N. D. 1982. Population dynamics of the wheat stem sawfly, Cephus cinctus (Hymenoptera: Cephidae), in wheat. Can. Entomol. 114: 755-88.

Kaya, O. 1976. The cereal leafminer, *Syringopais temperatella.* In: Anon (ed.) Proceedings of Symposium on Wheat Pests. CENTO Scientific Programme report no. 22, Ankara, pp. 49-50.

Koyuncu, N. 1976. Biology and control of Barytychius hordei. In: Anon (ed.) Proceedings of Symposium on Wheat Pests. CENTO Scientific Programme report no. 22, Ankara, pp. 29-35.

Lodos, N. 1973. Wheat pests and their importance in Turkey. In: Anon (ed.) Proceedings of Panel on Pests and Diseases of

Wheat. CENTO Council for Scientific Education and Research. Karaj, Iran. pp. 28-56.

Lodos, N. 1981a. Pentatomid pests of wheat in Turkey. Bulletin of the European Plant Protection Organization 11: 9-12.

Lodos, N. 1981b. Aelia speices and their importance in Turkey. Bulletin of the European Plant Protection Organization 11: 29-32.

Long, D. B. 1960. The wheat bulb fly, *Leptohylemyia coarctata* Fall. A review of current knowledge of its biology. Rothamstad Experimental Station Report for 1959, 216-229.

Lowe, H. J. B. 1982. Some observations on susceptibility and resistance of winter wheat to the aphid *Sitobion avenae* (F.) in Britain. Crop Protection 1: 431-440.

Metcalf, C. L., Flint, W. P. and Metcalf, R. L. 1962. Destructive and Useful Insects, their Habits and Control. McGraw Hill, New York. 1087 pp.

Paulian, F. and Popav., C. 1980. Sunn pest or cereal bug. In: Anon (ed.) Wheat. Documenta Ciba-Geigy, Basle. pp. 69-74.

Peterson, R. F. 1965. Wheat, Botany, Cultivation and Utilization. Leonard Hill, London. 422 pp.

Powell, W., Dean, G. J., Dewar, A. and Wilding, N. 1981. Towards integrated control of cereal aphids. Proceedings of the 1981 British Crop Protection Conference - Pests and Diseases 201-206.

Rabbinge, R. and Carter, N. 1984. Monitoring and forecasting of cereal aphids in the Netherlands: a subsystem of EPIPRE. In: Conway, G. R. (ed.) Pest and Pathogen Control: Strategic, Tactical and Policy Models. John Wiley & Sons, New York. pp. 242-53.

Rabbinge, R., Sinke, C. and Mantel, W. P. 1983. Yield loss due to cereal aphids and powdery mildew in winter wheat. Mededelingen van de Faculteit Landbouwhogeschool van de Rijksuniversiteit to Gent 48: 1159-67.

Rivnay, E. 1956. The biology and control of the Cereal Leaf Miner (*Syringopsis temperatella* Led.) in Israel. Ktavim 7:5-23.

Simmonds, N. W. 1976. Evolution of Crop Plants. Longman, London. 339 pp.

Starks, R. J., Muniappan, R. and Eikenbary, R. D. 1970. Interaction between plant resistance and parasitism against the greenbug on barley and sorghum. Annals of the Entomological Society of America 65: 650-655.

Sunderland, K. D., Chambers, R. J., Stacey, D. L. and Crook N. E. 1985. Invertebrate polyphagous predators and cereal aphids. In: Dedryver, C. A. (ed.) Integrated Control of Cereal Pests. Bulletin of the IOBC West Palaerctic Regional Section, 105-114.

Sunderland, K. D., Chambers, R. J. and Carter, O. C. R. (1988, in press). Potential interactions between varietal resistance and natural enemies in the control of cereal aphids. In: Cavallorao,

R. and Sunderland, K. D. (eds.) Integrated Crop Protection in Cereals. Balkema, Rotterdam.

Talhouk, A. M. 1969. Insects and mites injurious to crops in Middle East countries. Monographien zur angew. Ent. no. 21. Verlag Paul Parey, Hamburg. 239 pp.

Turnock, W. J. 1977. Adaptability and stability of insect pest populations in prairie agricultural ecosystems. Tech. Bull. Univ. of Minnesota Agric. Expt. Stat. no. 310, 89-101.

Van Emden, H. F. 1983. Pest management - routes and applications. Antenna 7: 163-178.

Vereijken, P. H. 1979. Feeding and multiplication of three cereal aphid species and their effect on yield of winter wheat. Centrum voor Landbouwpublikaties en Landbouwdocumentatie, Wageningen. 58 pp.

Vereijken, P. H. 1986. From conventional to integrated agriculture. Netherlands Journal of Agricultural Science 34: 387-393.

Vereijken, P. H., Edwards, C., El Titi, A., Fougeroux, A. and Way, M. J. 1985. Management of farming systems for integrated control. Bulletin of the International Organization for Biological Control, West Palaerctic Regional Section. 34 pp.

Vickerman, G. P. and Sunderland, K. D. 1975. Arthropods in cereal crops: Nocturnal activity, vertical distribution and aphid predation. J. appl. Ecol. 12: 755-66.

Way, M. J. 1978. Integrated pest control with special reference to orchard problems. Report of the East Malling Research Station for 1977. 187-204.

Way, M. J. and Cammell, M. E. 1979. Optomising cereal yields - the role of pest control. Proceedings of the 1979 British Crop Protection Conference - Pests and Diseases 663-672.

Zohary, D. 1969. The progenitors of wheat and barley in relation to domestication and agricultural dispersal in the Old World. In: Ucko, P. J. & Dimbleby, G. W. (eds.) The Domestication and Exploitation of Plants and Animals. Duckworth, London. 44-66.

10

Pecan Domestication and Pecan Arthropods

M. K. Harris

The pecan <u>Carya illinoensis</u> (Wang.) K. Koch (Fam. Juglandaceae) is a deciduous iteroparous tree indigenous to a large area of North America extending from southern Illinois into Mexico (Fig. 1). The earliest written accounts of pecans are provided by 17th century Spanish explorers who noted the high quality of the nuts, the irregularity of their bearing and their intensive use as food by the Indians (see Brison 1974). Early settlers, naturalists, scientists and others prized the nuts as well. George Washington's appetite for them remained intact despite his wooden dentures and Thomas Jefferson anticipated the massive eastward movement by more than a century by growing trees from nuts at Monticello.

PECAN BIOLOGY AND DOMESTICATION

The wind pollinated monoecious pecan reproduces from seed. The population is highly heterozygous and outcrossing is enhanced by protygyny, where the imperfect female flowers become receptive to pollen before catkins on the same tree have mature pollen, and protandry, where most catkins mature and release pollen before female flowers on the same tree are receptive. The first propagation of relatively identical trees in the middle of the 19th century is credited to the black Louisianan slave Antoine who demonstrated that vegetative propagation of scions of a preferred tree onto pecan rootstocks would produce a tree that bore nuts characteristic of the scion parent (Brison 1974). By the early 1900s, pecan orchards were being introduced throughout the southern U.S. east of the Mississippi where many survive to this day. Orchards

Figure 1. - Aboriginal distribution of pecan (Little 1971).

in this new area had also been established from nuts and more were propagated that way as well. Pecans produced from scions in this region kept the name of the parent tree, like 'Mahan' or 'Stuart', and those grown from nuts were called "seedlings".

Pecan production in the native range also fluorished at this time because, even though the nuts produced by the wild trees were smaller (averaging 200/kg vs. 130 or less/kg for named varieties), contained proportionally less kernel (45% vs. 50%), and had harder, thicker shells that were more difficult to crack, they were always tasty and found a ready market. The vegetative propagation of named varieties within the native range of pecan also occurred in parallel with eastern expansion, but at a much slower rate because of the competition of the "native" trees. Wild trees that have been thinned and cleared of competing trees and brush are referred to as natives. Native groves are amenable to management in much the same manner as orchard trees and the domestication process of converting wild pecan to a native status is still underway (Olcott-Reid and Reid 1986). Today, about 80% of the pecans produced in Texas and more than half of those produced in the indigenous range of the pecan come from trees nature planted (Harris 1983). Replacement of these trees with selected vegetatively propagated cultivars is slow because the trees are long-lived, certainly in excess of 100 years, their smaller nut meats are of higher quality and moderately priced, and acceptable if not equal in attractiveness on baked goods and candies where the smaller whole halves provide good eye appeal with less weight and volume.

Thus, the commercialization of pecan trees began in earnest in the latter part of the 19th and early 20th century. Pecan in the indigenous range was primarily developed by selective retention of irregularly spaced pecan trees, while clearing and thinning surrounding areas to provide additional light and nutrients. A grass understory also developed and served to graze cattle and provide a support for equipment like tractors and sprayers as more intensive pecan management practices arose. Cow/pecan operations are common, if not still the predominant form for pecan production in the native range. Vegetative reproduction of pecan allowed a gradual establishment of orchards in the native range and provided for the rapid eastward expansion of pecan orchards in Alabama, Florida, the Carolinas, and, most notably, Georgia. Westward expansion occurred a generation later in the Mesilla Valley of New Mexico and Texas, and in the 1950s in Arizona and Mexico, and is perhaps catching on in California today. Large foreign plantings occur in Australia, Brazil, South Africa, and Israel and are generally less than 50 years old. The pecan nut is popular with consumers everywhere and has begun to emerge from a provincial to a cosmopolitan status as more emphasis is placed on advertising and marketing world-wide.

The anachronistic features of pecan cultivation, particularly within the indigenous range, distinguish this agricultural plant from most others grown in the U.S. because the others are typically of exotic origin and consist of highly selected cultivars resulting from a much longer history of selective propagation by man. In addition, the pest complex associated with pecan in North America is comprised of arthropods that have coexisted with pecan for millenia, whereas exotic crops are typically beset by pests of diverse origins (Harris 1980). If one seeks to understand how plant domestication affects insect association with plants, then the pecan ecosystem is an ideal crucible for such investigation because a wide range of ecological situations from quite undisturbed to highly domesticated orchards, with many intermediate permutations, are available for study (Maggio et al. 1983). This opportunity does not exist with most crop plants because primitive and intermediate forms have been eschewed commercially and displaced from the land at an accelerating rate by improved cultivars or, more likely, by different and exotic crops (Harris 1980). The indigenous crops that once sustained native peoples are now grown most intensively on another continent. The pecan in its indigenous range is at an early stage of domestication and the arthropod pest complex consists of the same species that have been associated with it in the wild.

The need for understanding how plant domestication affects insect associations has not been generally perceived among 1) basic entomologists, because they typically study insect interactions with economically unimportant plants in order to further our knowledge of biology, and 2) applied entomologists, because they typically study exotic crops or insects in strictly cultivated contexts and emphasis is placed on obtaining a quick solution to pest problems that seem to proliferate faster than resources can be mustered to address them. The advent of integrated pest management (IPM) emphasizing factors like: "the ecosystem is the management unit"; "maximization of use of predators, parasitic, host plant resistance, weather and other natural controls"; and "considering an interdisciplinary systems approach essential to success" (Bottrell 1979), has alerted practitioners to the importance of a more holistic understanding in finding durable solutions to insect problems, and identified how shallow our knowledge base is in these areas. The pecan agroecosystem provides a readymade natural experimental setting to examine this subject.

PECAN MASTING

Pecans grow in alluvial soils adjacent to rivers and streams and canopies often cover 30% or more of presumably undisturbed areas (Harris et al. 1983). Reproduction is from seed and bearing typically commences 10 or more years after germination, although horticulturalists have selected precocious cultivars that initiate fruiting in 5 years or so (Brison 1974). Average tree longevity is unknown, but stems with more than 100 growth rings are common and some with over 200 have been found. Furthermore, orchards planted in the last century continue to bear large crops, although not regularly. Nut production, once fruiting begins, is irregular and appears to be relatively synchronous with other pecan in the region. Fig. 2 shows nut production records for Texas from 1919 - 1984. These features of pecan, viz., relatively dense stands extending over a large area, delayed maturity, long-lived individuals, and synchronous reproduction on an irregular basis, indicate the pecan exhibits masting. Masting is thought to be a reproductive mechanism that increases the chances of flowers surviving to dehiscence by satiating the few fruit predators that remain after the barren production interval and also increases the chances of mature fruit surviving to germinate later by satiating scatter/hoarders and other dispersal agents whose numbers are also disproportionately low in relation to the mast. Willson (1983) recently reviewed masting and noted that more information is needed on most aspects of the phenomenon before the attendant hypothesis can be fully tested.

The cause and effect relationships that establish masting in pecan are largely unknown. Primary factors probably are the genetics of pecan in response to weather,

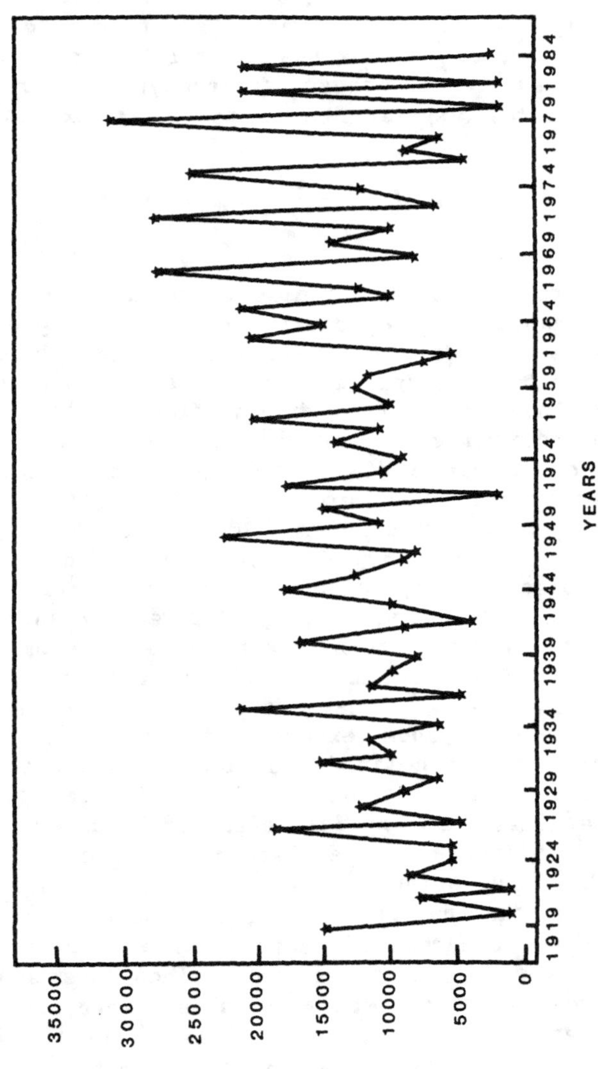

Figure 2. - Annual Pecan production in Texas from 1919-1984 (Murfield and Pratt 1981).

nutrient availability and pests. The annual production of large crops of pecans, i.e. 500 Kg/ha or higher, is generally restricted to young orchards for a decade or so after first coming into bearing and to well spaced intensively managed trees in small plots for short periods of time. Pecan orchards as they mature typically show yield patterns like those in Fig. 3; no yield the first few years after planting, followed by heavy annual bearing for a time and then the bane of the commercial producer -- biennial production. The latter appears to be due primarily to shading because temporary "cures" can be achieved by thinning or heavy pruning. An intriguing feature of the biennial bearing tendency of mature orchards and native groves is the synchrony of production or barrenness over wide areas. Orchards are planted and managed and groves are thinned by persons operating more or less independently, so that equal activity presumably occurs in even numbered and odd numbered years, yet relative synchrony of nut production over wide areas ultimately prevails (see Fig. 2). This indicates a strong intrinsic component, probably genetic, is at work that resets the clock in each tree. Masting, although as yet poorly understood, is clearly a strong force in shaping pecan biology and will be considered central to understanding how pecan domestication affects and is affected by arthropods.

PECAN ARTHROPOD COMPLEX

The pecan arthropod pest complex consists of several dozen phytophagous species (Fig. 4). The integrated pest management of this complex has been extensively examined in recent reviews (Cooper et al. 1983, Harris 1983, 1985a) and intensive studies on foliage and nut feeding pests (Ring et al. 1985, Harris et al.1986) show that the key arthropod pests are 2 species of nut feeders. Nevertheless, foliar feeders like aphids, pecan phylloxera, mites, walnut caterpillar, grasshoppers and others do occasionally cause epidemics and adequate data are available to allow some conclusions to be drawn from these interactions.

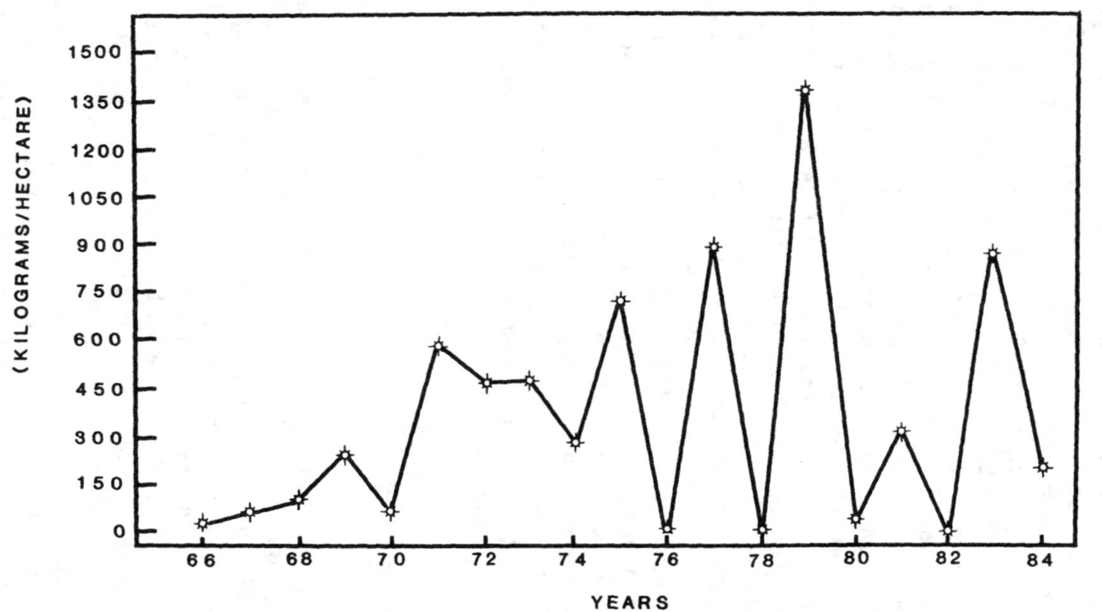

Figure 3. — Annual pecan production in an eight hectare pecan orchard planted in 1956 near College Station, Texas (Courtesy Dr. J. Benton Storey, Horticulturalist, Texas A&M University).

Figure 4. This seasonal pecan pest profile indicates some possible arthropod problems associated with various developmental stages of the pecan and shows when management may be required.

PESTS: PhSc

LT+D YA--------------------------------BA-------------
Ca WC$_1$ WC$_2$ WC$_3$
Sa M---
Lc WW$_1$ WW$_2$
Ph S-------------
C$_1$ C$_2$ C$_3$
Sp T
W-------------

PLANT STAGE: D   Bb   Po   WS   DS   SS   LD
MONTHS: J F M A M J J A S O N D

PESTS: Ph is <u>Phylloxera</u> spp., Sc is <u>Melanaspis</u> sp., Lt + D is leaf tatterers and defoliators, Ca is <u>Catocala</u> spp., Sa is <u>Periclista</u> spp., LC is <u>Acrobasis juglandis</u>, Sp is <u>Clastoptera</u> spp., YA is <u>Monellia</u> sp. and <u>Monelliopsis</u> sp., WC is <u>Datana integerrima</u>, M is <u>Eotetranychus hicoriae</u>, WW is <u>Hyphantria cunea</u>, C is <u>Acrobasis nuxvorella</u>, S is <u>Cydia caryana</u>, W is <u>Curculio caryae</u> and T is <u>oncideres</u> sp. (The position aligns with when controls may be needed). PLANT STAGE: D is dormancy, Bb is budbreak, Po is pollination, WS is liquid endosperm stage of nut development, DS is kernel formation in nut, SS is dehisence of involucre surrounding nut and LD is leaf dehisence.

ARTHROPOD FOLIAGE FEEDERS

A pecan leaf life table study comparing insecticide managed and unmanaged, orchard, native and wild pecans, concluded that pecan foliage was typically reduced across all plots by less than 20% during each season by arthropods and that 2-3 insecticide treatments during the season may have extended leaf retention in the managed plots by a week or two in the fall compared to the unmanaged plots (Ring et al. 1985). Foliage feeders appeared to pose little threat to pecan; yet, selective studies of many pecan/arthropod interactions had clearly shown that severe or even complete defoliation could occur during the growing season due to a number of arthropods. Severe defoliation epidemics can apparently be ascribed to three sets of circumstances, 1) natural, 2) chemically induced, and 3) genetically induced, with the latter two due to mans activities.

Natural Epidemics

The walnut caterpillar in August 1973 defoliated juglandacious trees over tens of thousands of square miles in Texas (Harris et al. 1982). Previous reports indicate such widespread defoliations occur several times each century and presumably these periodic defoliations extend back in time for thousands of years. This interaction unquestionably reduces the ability of the affected trees to produce nuts because the total loss of leaves in mid-season eliminates the opportunity for carbohydrate formation through photosynthesis until secondary buds put out new leaves. Evidence gathered in the 1973 epidemic shows that ca. 10 weeks were required for inducing buds and growing fully expanded new leaves (Harris et al. 1976). The remaining growing season was short (4-6 weeks) for the trees to recoup the carbohydrate reserves used to establish the new canopies let alone achieve a surplus. Previous studies have shown defoliation in August has adverse effects on future nut production (Worley 1971). The refoliated trees did retain their foliage for a month or so longer than did insecticide treated (and thus undefoliated) trees. This extension of the photosynthetic period probably did allow for the previously defoliated trees to recover lost energy reserves, but it also exposed them to an increased risk of freeze damage.

The effect on the walnut caterpillar was also devastating because oviposition occurs on the bottom surface of mature leaves and the second summer generation females had to emigrate from the defoliated areas to find suitable oviposition sites (Harris et al. 1982).

This scenario is somewhat troublesome for a scientist who expects an interspecific regularly occurring interaction in a natural system to ultimately reach some kind of relatively stable equilibrium that optimizes resource exploitation while at least minimizing mutually disruptive interactions. Masting holds the key to understanding this apparently chaotic and enigmatic situation. Although the walnut caterpillar epidemic lowers the efficiency with which nuts are produced by the tree population, the masting cycle is preserved if not enhanced by the widespread epidemic. Renegade trees would have to possess heritable qualities for resistance and the ability to either 1) sequester their additional carbohydrate reserves until the next masting cycle or 2) protect their developing nuts and seedlings produced during otherwise barren periods from attack. Resistance to walnut caterpillar, or any periodic defoliation for that matter, could arise from a single gene. But, simultaneous realignment with the masting cycle or development of resistance to nut and seedling feeders would most likely require many additional and concurrent gene changes. Such macromutations are unlikely (see Fisher 1958) and this probably explains why the present system is preserved.

Chemically Induced Epidemics

Pecan aphids, mites and leafminers are notorious defoliators on occasion in intensively managed systems but rarely if ever observed to cause widespread epidemics in wild trees (Boethel 1981, Harris 1982, Tedders 1978). Extensive research indicates that these phytophages are primarily kept at low levels by their natural enemies and to a lesser extent by the less nutritious and more diverse environment provided by the wild trees (Harris 1983, Liao et al. 1984). The regular use of broad spectrum pesticides severely reduces populations of the natural enemies and resistance to the chemicals develops faster and to higher levels in the phytophages so that epidemics result.

218

Genetically Induced Epidemics

The pecan phylloxera complex (Harris 1982, Stoetzel and Tedders 1981) is known to contain four species and _Phylloxera devastatrix_ Perg. is the principal pest. Damage results from golf ball sized galls that form around stem mothers that settle on developing foliage in the spring. By June, the affected foliage senesces and in severe cases the entire tree can be more than 50% defoliated. Epidemics in managed orchards typically are limited to outbreaks on a single cultivar and while every tree of that particular clone may be severely affected, neighboring trees of different clones will have few if any galls even when branches intertwine with the susceptible clone. Wild trees are rarely found with severe galling and I have never found even two adjacent wild trees to be severely affected. The susceptible cultivars are typically the popular older selections that have been grown in the same area for many decades like 'Texas Prolific', 'Success' and 'Stuart'.

Presumably, the successful establishment of a gall requires a suitable synchrony between the hatching phylloxeran and development of the foliage, and a genetic match between phytophage and host. The latter factor is most familiar to plant pathologists with systems like rusts and small grains (Browning 1974) and studies of black pine scale on pines (Edmunds and Alstadt 1978) show it occurs in insects as well. Epidemics of pecan phylloxera do not occur in wild pecan because opportunities for initial gall establishment are limited due first to incompatibilities at the genetic level and secondly to vicissitudes of natural enemies and weather. The maintenance of an established infestation may also be tenuous in the wild ecosytem because local eradication from the infested tree essentially reestablishes the original and rather formidable barriers to infestation.

The impact of reducing the diversity of pecans from millions of genetically different trees to a few widely grown cultivars is poorly understood. The genetic predictability that results from widespread propagation of popular cultivars affects disease as well (Cole 1957) and apparently pecan domestication has increased the vulnerability of the tree to a number of pests.

ARTHROPOD NUT FEEDERS

Numerous arthropods feed on pecan reproductive structures during their development (Harris 1983). There appear to be two species of primary importance. These are the pecan nut casebearer, <u>Acrobasis nuxvorella</u> Nuenzig (Lepidoptera:Pyralidae), and the pecan weevil, <u>Curculio caryae</u> (Horn) (Coleoptera: Curculionidae).

Pecan Nut Casebearer

The monophagous multivoltine pecan nut casebearer (PNC) overwinters as an early instar larva in a hibernaculum attached to a pecan bud. Larvae of the overwintering generation begin feeding when pecan bud growth begins in spring and tunnel within developing shoots to complete development. PNC adult females emerge from pupae when pecan flowers have been pollinated and deposit the first summer generation eggs on the small nutlets. Larvae hatch and each destroys about two nutlets while completing larval development within about two weeks; the infestation typically spans 3 weeks (Ring and Harris 1984). Subsequent summer generations occur at about 6 week intervals with as many as four having been reported (Bilsing 1926, 1927). These too feed on developing nuts if they are available. In the absence of nuts, larvae typically spin a hibernaculum shortly after hatching. Bilsing (1927) reported some summer generation larvae complete development by feeding on buds and leaves, although most die in the attempt. The PNC is also a host for dozens of parasitic species (Gunasena 1985) and the first summer generation larvae successfully infesting nutlets are often highly parasitized (>20%). Pecan masting and PNC parasitization appear to act in concert to suppress PNC populations to inconsequential levels in mast years. A pecan nut life table study (Harris et al. 1986) showed high levels of PNC damage generally occurred only in the years of low flower production and typically was limited to the first summer generation.

Domestication of pecan affects interactions with PNC in primarily two ways. The increased availability of pecan nutlets in most years means the PNC is no longer limited by food, and intervention with broad spectrum pesticides interferes with the natural enemies of the PNC, thereby diminishing their role in regulating the pest.

The solution currently being implemented involves limiting the use of insecticide early in the season to occasions when a target pest threatens real damage to the crop (Harris 1985a). A degree-day model has been developed to anticipate infestation by PNC (Ring et al. 1983, Ring and Harris 1983) and a sequential sampling plan has been devised to detect PNC densities greater or less than sufficient to damage 10% of the nutlets with insecticide only being used when damage greater than 10% is anticipated (Ring et al., unpublished). These approaches preserve natural enemies as much as possible and capitalizes on the escape from heavy infestation nature provides in mast years.

Pecan Weevil

The pecan weevil is a late season obligatory nut feeder of <u>Carya</u> species whose 2-3 year life cycle is ideally suited for developing high infestations in domesticated pecans (Harris 1985b, Harris et al. 1980). Adults typically emerge in August/September, mate, and eggs are laid in excavations made within the pecan kernel by the female's slender proboscis. Female longevity is ca. 25 days postemergence and she oviposits in ca. 25 nuts. Nuts that have developed to the gel stage (endosperm has begun to form distally) are susceptible for oviposition until shuck split. The tiny hole piercing shuck and shell of the pecan is sealed as nut growth proceeds and after developing for ca. 6 weeks within the protected interior of the nut, three larvae typically emerge through a single emergence hole and enter the soil to ca. 10 cm; each constructs a cell where they remain for 10 months, pupate, and then remain another year before emerging as an adult to repeat the cycle. About 10% of the larvae remain 22 months before pupation, which provides the 3 year life cycle.

The pecan weevil-pecan interaction is greatly influenced across years by the fluctuating, yet synchronous, patterns of nut production. Field studies indicate the pecan weevil population increases ca. three times from one generation to the next when unlimited food is available (Harris 1985b). Pecan production data from Texas (Fig. 2) shows state yields can vary 10-fold among years and this variation is accentuated even more at the county level and, perhaps more pertinently, in the few km^2 range accessible to a given pecan weevil population through natural dispersal. We are in the process of formalizing this argument (Harris and coworkers, unpublished), but given an unpredictable 10-fold variation in the annual nut crop and a 3-fold realized capacity for increase by the pecan weevil on a biennial basis, the preliminary calculations we have made indicate that large crops escape major depredation by pecan weevil because the insect populations tend to be low due to limited food supplies in earlier years.

The pecan weevil-pecan interaction is also influenced within years by variations in nut phenology in the pecan population. Nut phenology is relatively uniform within a tree and the onset of susceptibility to oviposition occurs as early as mid August until as late as early October. The bulk of pecan weevil emergence occurs between Aug. 20 and Sep. 10, so that most weevils have access to most pecans, although pecans entering gel stage after mid September would, on average, escape pressure from weevils that emerged 25 days or more earlier, since those weevils would have already completed their life span (Harris 1976).

Drought creates another factor that can realign the phenology of the interaction in soils that require more than 60 Kg/cm^2 in order to be penetrated (Blanchard 1981). Significant numbers of weevils are delayed from emerging from such drought hardened soils until rain or irrigation softens the soil below this threshold impedance level. This can result in peak emergences as late as the latter part of October when many pecan varieties have already undergone shuck split and are no longer susceptible to attack, leaving late maturing varieties to absorb the attack (Harris and Ring 1980). The onset and duration of drought is unpredictable, so that different trees in a population can be attacked in different years.

Pecan domestication has provided the pecan weevil a regular food supply and populations tend to increase to high levels. The natural state of the pecan weevil appears to be one of persisting at a low level in a food limited environment and exploiting opportunities afforded in mast years. The long life cycle, dispersed and quiescent subterranean condition and generally low populations of pecan weevil probably account for the fact it has few natural enemies and none appear to be effective in preventing epidemics in domesticated pecans (Harp and Van Cleave 1976, Harris and Ring 1980). Current management practices protect commercial crops from damaging weevil populations by monitoring adult emergence from the soil and treating them with insecticide before oviposition can occur in susceptible nuts. Generally, 2 - 3 treatments are needed for native/seedling pecans and fewer may be effective where weevil emergence is asynchronous with nut phenology of particular pecan varieties planted in independently accessible blocks in certain years (Harris et al. 1980).

CONCLUSIONS

Pecan domestication is producing an environment where nuts are available in large numbers in most years from a few cultivars compared to wild pecan environments where masting by a genetically diverse tree population is the rule. The monophagous pecan nut feeding arthropods are released from their previously food limited state and populations increase to assume a new equilibrium level determined more by natural enemies and climatic factors. Insecticide intervention is needed to preserve commercial pecans when arthropod populations increase to unacceptable levels, but this practice reduces the beneficial role of natural enemies on target pests and nontarget pests alike. By carefully planning before the orchard is planted, varieties with desireable phenologies can be planted in separately treatable blocks that still allow wind pollination and pest monitoring systems to be implemented to ensure that insecticide use is minimized (Harris 1981). The defensive mechanisms we have inherited from nature can be of greater value in the profit driven environment of commercial pecan production, if we develop a better understanding of them and how they can be exploited.

REFERENCES CITED

Bilsing, S. W. 1926. The life history and control of the pecan nut casebearer (_Acrobasis_ _caryae_). Tx. Agric. Exp. Stn. Bull. 328. 77p.

Bilsing, S. W. 1927. Studies on the biology of the pecan nut casebearer. Tx. Agric. Exp. Sta. Bull. 347. 71p.

Blanchard, C. E. 1981. Emergence of the adult pecan weevil in relation to soil mechanical impedence and moisture. M. S. Thesis. Texas A&M University, College Station, TX.

Boethel, D. J. 1981. Resistance in the pecan leaf scorch mite: implications in pecan pest management. Misc. Publ. Entomol. Soc. Amer. 12: 31-44.

Bottrell, D. R. 1979. Integrated pest management. Counc. on Environ. Qual. U.S. Govt. Printing Office, Wash., D.C. 120p.

Brison, Fred R. 1974. Pecan Culture. Capital Printing, Austin, Tex.

Browning, J. A. 1974. Relevance of knowledge about natural ecosystems to development of pest management programs for agroecosystems. Proc. Am. Phytopathol. Soc. 1: 191-199.

Cole, J. R. 1957. Will scab eventually affect all varieties of pecan? Proc. SE Pecan Grow. Assoc. 50: 79-81.

Cooper, J. N., G. R. McEachern, G. M. McWhorter, J. D. Johnson and M. K. Harris. 1983. Implementation of pest management in Texas pecan production. In Pecan Pest Management - Are we there? J. A. Payne, ed. pp.111-120. Entomol. Soc. Amer. Misc. Pub. 13 (2).

Edmunds, G. F. Jr., and D. N. Alstad. 1978. Coevolution in insect herbivores and conifers. Science 199: 941-946.

Fisher, R. A. 1958. The genetic theory of natural selection. Dover, New York.

Gunasena, G. H. 1985. Some factors affecting hickory shuckworm diapause termination and parasites of hickory shuckworm and pecan nut casebearer. M.S. Thesis. Texas A&M Univ., College Station, Tex. 65p.

Harp, S. J. and H. W. Van Cleave. 1976. Biology of the pecan weevil. Southw. Entomol. 7:21-39.

Harris, M. K. 1976. Pecan weevil adult emergence, onset of oviposition and larval emergence as affected by the phenology of the pecan. J. Econ. Entomol. 69:167-170.

Harris, M. K. 1980. Arthropod-plant interaction related to agriculture emphasizing host plant resistance to arthropods and pathogens in agricultural plants. Agric. Comm., Tex A&M Univ. MP 1451.

Harris, M. K. 1981. Planning pecan weevil management before planting the orchard. Pecan Quart. 15:24-26.

Harris, M. K. 1982. Genes for resistance to insects in agriculture with a discussion of host-parasite interactions in Carya. Proc. Workshop Genet. Host-Parasite Interactions in For. Pudoc, Wagening.

Harris, M. K. 1983. Integrated pest management of pecans. Ann. Rev. Entomol. 28:291-318.

Harris, M. K.   1985a.   Understanding pecan IPM.   Pecan South.

Harris, M. K.   1985b.   Pecan phenology and pecan weevil biology and management.   In Pecan Weevil: Research Perspective.   W. W. Neel ed., pp.   51-58.   Quail Ridge Press, Brandon, Miss.   127p.

Harris, M. K., B. L. Cutler, and D. L. Ring.   1986.   Pecan nut loss from pollination to harvest.   J.   Econ. Entomol. 79:1653-1657.

Harris, M. K. and D. R. Ring.   1980.   Adult pecan weevil emergence related to soil moisture.   J. Econ. Entomol. 73:339-43.

Harris, M. K., W. G. Hart, M. R. Davis, S. J. Ingle and H. W. Van Cleave.   1976.   Pecan identification, defoliation, refoliation, and nut yield in relation to walnut carterpillar attack on native pecans.   Pecan Q. 10:12-18.

Harris, M. K., W. G. Hart, M. R. Davis, S. J. Ingle, R. C. Maggio and R. D. Baker.   1983.   Remote Sensing of Pecan.   In Pecan Pest Management - Are We There?   J.A. Payne, ed., pp 77-88.   Misc. Pub. Entomol. Soc. Amer. 13:Misc. Pub.

Harris, M. K., R. C. Maggio, W. G. Hart, S. J. Ingle and M. R. Davis.   1982.   Use of remote sensing to characterize defoliation of pecans by the walnut caterpillar.   Southw. Entomol.   7:146-154.

Harris, M. K., D. R. Ring, B. L. Cutler, C. W. Neeb and J. A. Jackman.   1980.   Pecan Weevil Management in Texas. Pecan 14:3-14.

Liao, H. T., M. K. Harris, F. L. Gilstrap, D. A. Dean, C. W. Agnew, G. J. Michaels and F. Mansour.   1984. Natural enemies and other factors affecting seasonal abundance of the blackmargined aphid on pecan. Southw. Entomol. 9:404-420.

Little, E. L. 1971.   Atlas of United States tress.   Vol. 1.   Conifers and important hardwoods.   U.S.D.A. For. Serv. Misc. Publ. 1146.

Maggio, R. C., R. D. Baker and M. K. Harris.   1983.   A geographic data base for Texas pecan.   Photogrammetric Engineering and Remote Sensing. 49:47-52.

Murfield, D. and W. L. Pratt.   1981.   Texas historic crops statistics.   Tex.   Crop   and   Livestock   Reporting Service. Tex. Dept. Agric.   Bull. 129.   Austin, TX.

Olcott-Reid, B. and W. Reid.   1986.   Profit from native pecans.   Am.   Fruit Grow.   106:16-17.

Ring, D. R., V. R. Calcote and M. K. Harris.   1983. Verification and generalization of a degree-day model predicting pecan nut casebearer (Lepidoptera: Pyralidae) activity. Environ. Entomol. 12:487-489.

Ring, D. R. and M. K. Harris. 1983. Predicting pecan nut casebearer (Lepidoptera: Pyralidae) activity at College Station, Texas. Environ. Entomol. 12:482-486.

Ring, D. R. and M. K. Harris. 1984. Nut entry by 1st generation pecan nut casebearer. Southw. Entomol. 9:13-21.

Ring, D. R., M. K. Harris and R. Olszak. 1985. Life tables for pecan leaves in Texas. J. Econ. Entomol. 78:888-894.

Stoetzel, M. B. and W. L. Tedders. 1981. Investigation of two species of _Phylloxera_ on pecan in Georgia. J. Ga. Entomol. Soc. 16:144-150.

Tedders, W. L. 1978. Important biological and morphological characteristics of the foliar-feeding aphids of pecan. U. S. Dept. Agric. Tech. Bull. 1579.

Willson, M. F. 1983. Plant reproductive ecology. John Wiley & Sons. New York.

Worley, R. E. 1971. Effects of defoliation data on yield, quality, nutlet size, and foliage regrowth for pecan. HortScience 6:446-447.

Subject Index